Sunlight to Electricity

Photovoltaic Technology and Business Prospects
Second Edition

Joseph A. Merrigan

The MIT Press
Cambridge, Massachusetts
London, England

This book was set in Times Roman by Graphic Composition, Inc., and printed and bound in the United States of America.

Library of Congress Cataloging in Publication Data

Merrigan, Joseph A.
    Sunlight to electricity.

    Bibliography: p.
    Includes index.
    1. Solar cells. 2. Photovoltaic power generation. 3. Solar energy. I. Title.
TK2960.M47   1982   621.31'244   81–18851
ISBN 0–262–13174–9   AACR2

Contents

3

Principles of Photovoltaic Energy Conversion
50

4

State of the Art in Photovoltaic Conversion Technology
68

5

Projected Developments in Photovoltaic Solar Conversion Systems
134

6

Business Opportunities in Photovoltaic Energy Conversion Systems
159

List of Figures and Tables

Figures

duce water internally and dc electrical power through an external circuit

---

Tables

Preface

This book is a 1980 assessment of the economic and technological prospects for development of solar cells into commercially feasible converters of sunlight to electrical energy. It is basically a revision of a book on this subject published by the MIT Press in 1975: *Sunlight to Electricity: Prospects for Solar Energy Conversion by Photovoltaics*. Much has happened since 1975 in public awareness of energy issues, price of energy, governmental policies toward energy development, and technical progress in solar cell manufacture. An updated study of the status of solar cell development and their prospects for economic feasibility in various markets is pertinent as we enter the decade of the eighties.

This new look at the prospects for solar energy use by photovoltaics reveals many changes in the factors that influence the growth of the solar cell business. Shortages, large price increases, and conservation of energy have led to forecasts of energy demand that are considerably lower than those made in the early 1970s. Many new governmental policies and industrial incentives have developed that are having a major impact on the development of alternative energy sources including solar cells. Research during the last seven years has made feasible many new technical approaches for making solar cells less costly and more efficient. Finally, there is now a six-to-seven-year business history of the terrestrial use of solar cells for electricity production. Based upon the dominating technical, market, and economic factors, a new forecast is made for the development of the solar cell business into the twenty-first century.

The first chapter is devoted to defining the probable US energy demand and supply until the year 2000. Included is an assessment of the demand for electricity—the output of solar cells. This chapter illustrates the long-term incentive to utilize solar energy. The extent of solar energy falling on the United States in relation to energy demands is treated in chapter 2, along with a description of the spectral quality of sunlight. Chapters 3 and 4 deal with the technology and state of the art in direct conversion of sunlight to electricity. The remaining chapters treat the economics of photoelectricity generation, market considerations and projections, and technological and business forecasts to the end of this century.

Sunlight to Electricity

1

Energy Use in the United States

Primary Energy Demand

One of the most important factors in the economic development of the United States is the availability of an adequate supply of power to maintain and improve its mechanized society. Conveniently usable energy is the result of many fossil fuel resources, the ability to convert these resources into easily distributed and usable forms, and a willingness to invest in development of these resources. In the early history of the United States, the readily available and broadly utilized source of energy was wood. In the last half of the nineteenth century, a more concentrated form of energy, coal, was used to fuel the industrial revolution. During the first half of the twentieth century, still another form of energy, oil, easier to distribute and cleaner to burn, was used in advancing a revolution in transportation. The relative positions of primary energy resources in supplying US energy over the past years are illustrated in figure 1.1.[1]

The average annual growth in energy use since 1850 has been 2.8%. The use doubled every 25 yr. This trend is shown in figure 1.2.[2] During the post-World War II era until the oil embargo of 1973, there was great economic development, and the price of energy became less and less expensive relative to the value of currency and things it could buy. Because

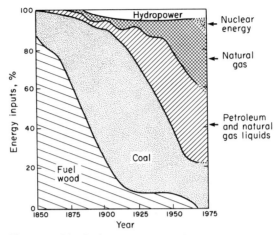

Figure 1.1 Distribution among energy inputs in the United States since 1850. (Based chiefly on data from reference 1)

of this and a multitude of other factors, the growth rate in energy consumption was inordinately high. Over the decade 1960–1970, use of energy in the United States rose from 45 × 10[15] to 68 × 10[15] Btu*/yr, an increase of 4.3% compounded annually.[3] Early in the 1970s, extensions of this trend with modifications for, among other things, population projections, gross national product projections, energy price and supply projections led to several reputable predictions of continued growth of the same magnitude.[3-9] These predictions are summarized in reference 10: the highest projection of annual growth rate, 4.2%, would result in primary energy consumption of about 240 × 10[15] Btu in the year 2000; the lowest projection, about 3%, in consumption of about 170 × 10[15] Btu in 2000.

---

*A British thermal unit (Btu) is the energy required to raise the temperature of 1 lb of water 1°F (Fahrenheit).

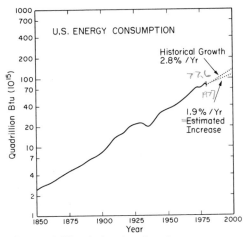

Figure 1.2 Historical and projected energy consumption in the United States. (Based chiefly on data from reference 2)

During the 1970s, many factors militated against such a high growth rate in energy use. Various environmental controls restricted the amount of noxious gases and particulates that could be released into the atmosphere. Because the burning of primary fuels, such as oil and coal, produces a certain effluent of gas and particulates, one of the easiest ways to stay within effluent guidelines was to burn less for the intended function. Automobile suppliers produced more fuel-efficient models. Industry conserved energy to reduce costs as well as to postpone construction of new power plants until new effluent-scrubbing techniques could be developed. Nuclear plant sites were restricted considerably. A national debate over the cleanness of the atmosphere and its relation to energy consumption reversed the prior decade's trend toward more pollution. The factor with perhaps the greatest impact on energy consumption was the rise in the price of crude oil. The Organization of Petroleum Exporting Communities (OPEC) exerted a new-found strength in 1973–1974 during the oil embargo and succeeded in setting the general price level of oil through 1980. The major role of OPEC in supplying oil to the world and a shortage of oil in the absence of OPEC production portend a long period of high-priced oil. The price of crude oil since 1965 is illustrated in figure 1.3

Because about 45% of the US energy demand is met by burning oil, this price trend and some spot shortages during the 1970s profoundly changed the public attitude toward conserving energy. A regulation was promulgated by the federal government in the late 1970s relative to how high the room temperature in public and business places could be in the winter and how low in the summer. A national speed limit of 55 mph was set. Maladjustments in gasoline allocations between regions caused long waits at automobile

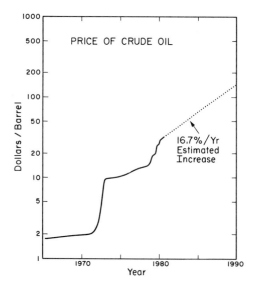

Figure 1.3 Historical and projected prices of crude oil.

service stations. Vacation travel was curtailed. Business was influenced considerably by transportation to the place of business. The public demand for automobiles changed from large to small, fuel-efficient cars. The long-standing trend toward larger luxurious cars that developed from 1950 to 1973 was suddenly reversed, causing a major impact on the automobile industry. From 1970 through 1979, US energy consumption increased an average of only 1.8%/yr to 79×10^{15} Btu. This use decreased to about 76×10^{15} Btu in 1980.[11]

Because of the recent national response to increasing energy costs, shortages, and pollution caused by burning primary fuels, current projections of energy demand through the rest of the century are much more conservative than those made in the early seventies. Gulf Oil Corporation predicted in 1980 that the rate of increase will be 2% annually through the 1980s, and about the same in the 1990s.[12] J. H. Krenz has made a very compelling argument that the rate of growth will be about 1.9%[13] There appears to be an increasing efficiency of energy use in the United States. In the 1950s, energy use per person increased by 1.71%/yr. In the 1960s, it increased 1.26%/yr, and in the 1970s, it increased less than 1%/yr. Over the last 100 years, the energy consumption per person has increased an average of 1.1%/yr. Also, a plot of the energy used per dollar of real gross national product against time shows that it has decreased over the last 60 years.[14] Population projections for the United States yield annual growth rates of 0.55–1.18% up to the year 2000.[15] If a 0.8%/yr population growth prevails, and we use the 100-year average annual growth in energy use per person (1.1%/yr), a 1.9% annual increase in energy consumption may be predicted. This projection is plotted in figure 1.2 along with an extension of the historical long-term trend in energy consumption. The lower projection yields a demand for 110 $\times$ 10^{15} Btu in the year 2000 compared to about 76 $\times$ 10^{15} in 1980.

Projections of energy use over a relatively brief period, 1980–2000, must be considered speculative at best. Many factors can affect what will actually occur. Under normal conditions, the most significant long-range determinants of energy demand are considered to be (a) economic activity (GNP), (b) cost of energy, (c) population, and (d) environmental controls. These have been found to explain most of the past changes in demand.[5] Factors that might be very important, such as supply limitations, political decisions, and major technical breakthroughs, are not reflected in the projections shown in figure 1.2. Current predictions call for an increase in GNP of 2–2.5%/yr per person and less than 1%/yr growth in population. Both are somewhat smaller than in the past. This would portend a decrease in the energy demand growth rate relative to that in the last three decades. Considerably higher energy costs and increased consideration of the impact of pollution also militate against a high rate of growth in energy consumption.

The demand for energy can be subdivided into areas of major use for a better insight into future needs. Figure 1.4[3] gives a good perception of the flows of energy through US society. Primary energy sources—coal, oil, gas, nuclear, hydro, and geothermal—are used to make a secondary energy source, electricity, which, together with the primary fuels, supplies energy to the consuming sectors: residential and commercial; industrial; transportation; nonenergy petrochemicals, fertilizers, and plastics; and so forth. The units in figure 1.4 are relative to 100% of the energy used in the United States in 1970. About 22.2% of the energy went into residential and commercial uses, predominantly for conditioning the indoor atmosphere; 29.3% into industry; 22.8% into transportation; 5.62%, into nonenergy; and 21% into

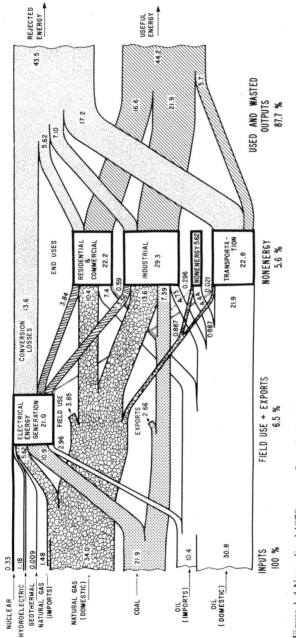

Figure 1.4 Normalized 1970 energy flows in the United States. (Data from reference 3)

generation of electricity. It is interesting to note that of all the primary input, only half provides useful work, while the other half is wasted as byproduct energy rejected to the environment. Transportation and generation of electricity are very wasteful. About 75% of the fuel that goes into transportation is wasted. Of the primary energy that goes into producing electricity, a very convenient form of energy, 65% is lost. By the time the electricity is utilized and partially lost through inefficiencies, the figure is even higher. Generation of electricity is expected to grow at about double the rate of growth of total energy demand.[16] This is largely due to the relative ease of distribution of electricity to residential, commercial and industrial sectors; its convenience; and its use in pollution abatement. The substitutability of primary fuels in the production of electricity also makes it probable that electricity may be an increasingly utilized form of end use energy. Consumers would not have to change their facilities from, say, oil burning to coal burning, as the relative costs and supplies of primary fuels changed. Such changes would be left to the utility companies to handle, perhaps with multifueled generators. At present, nearly 30% of the primary energy used in the United States goes into generation of electricity. Slightly over half of the electricity is used in the residential and commercial sector, with the rest going predominantly into industry.

The distribution of energy among the three main categories of end uses in the United States in 1978 is illustrated in figure 1.5.[17] The Exxon Company[18] expects that the average growth rate in energy use through the next decade will be greater for the residential and commercial sector than for the industrial and agricultural sector, which in turn will be higher than for the transportation sector. Figure 1.6 illustrates energy uses by sector as projected by the Federal Energy Administration in 1975.[19] Although the total growth in

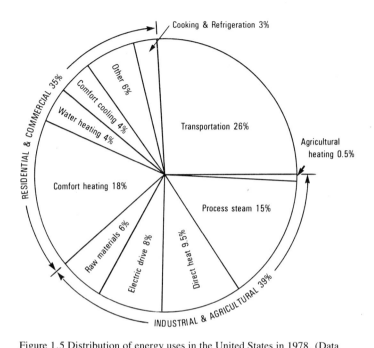

Figure 1.5 Distribution of energy uses in the United States in 1978. (Data from reference 17)

energy demand may not grow at the rate projected in 1975, before the major energy price increases of 1979, the relative demands between consuming sectors will probably be much as depicted.

Increases of efficiency in use of primary fuel in the transportation sector, which will grow relatively slowly, may be offset somewhat by the increased use of electricity, an inefficient use of primary fuels, in the more rapidly growing sectors. Hence, one might predict that the past trends in overall energy efficiency, reflected in the ratio of energy used to produce work to energy rejected to the environment, would remain about the same—about 50% (figure 1.4). However, it has become evident recently that the United

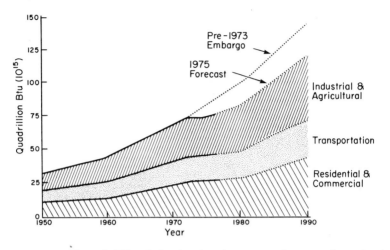

Figure 1.6 Historical and projected energy uses by consuming sectors in the United States. (Data from reference 19)

States has been reducing the energy intensity of the economy.[20] This can occur because of energy efficiency improvements, energy conservation, and changes in the mix of production. In 1978, 6% less energy was required per unit of GNP than the pre-1970 average. Further reductions are expected in all using sectors. Conservation and increased efficiency of heating and cooling in the commercial and residential sector will be accomplished by automatic thermostats and more insulation. The industrial sector is expected to continue increases in energy efficiency by multiple use of energy, for example, by using the process steam produced in electricity generation to provide space heating. The efficiency of transportation will increase by paying more attention to, for example, engine efficiency, weight of vehicles, and mass of vehicle per passenger. By the year 2000, 30% less energy is expected to be required per unit of GNP than was required between 1960 and 1975.[20] This national reaction to energy prices and the variety of other factors that affect the demand for energy lead to much more conservative estimates of energy demand than were prevalent a decade ago. From information available up to October 1979 (before some major oil price increases—figure 1.3), Exxon Company[20] predicted a 0.8% average annual growth in US energy demand from 1978 to 1990 and a 1.6% annual growth rate during the 1990s.

Primary Energy Supply

The world supplies of coal, oil, gas, wood, oil shale, tar sands, and uranium are theoretically adequate for centuries. However, they are not evenly distributed across the globe. They often pose difficult economic, technological, and environmental problems, and they vary greatly in their accessibility for use. The Global 2000 Study's projections indicate increasing problems rather than relief in supplying enough energy for projected world demand well into the twenty-first century.[21] About half of the world's energy usage is supplied by oil.[21,22] Petroleum production capacity is not increasing as rapidly as demand. A peak in oil production is expected before the end of the twentieth century. A transition away from oil dependence will probably occur in the next 10–40 yr. The direction of transition is not clear. Worldwide, it may be toward nuclear energy, coal, possibly gas, or perhaps solar energy. In the 200-yr history of the United States, development of energy supplies was left pretty much to the natural workings of free enterprise. The recently recognized importance of future supplies in the socioeconomic life of the nation has led to establishment of a cabinet-level Department of Energy to help ensure that national policies are established and appropriately implemented. World competition for the most convenient, least expensive, and cleanest energy sources is increasing yearly.

In the United States, about 45% of the annual energy demand is met by oil. Natural gas accounts for about 26%; coal, 21%; nuclear, 4%; and hydro and others, 4%. About 40% of the oil is imported. Slightly over three fourths of the total energy supply is produced domestically. A trend toward increasing dependence on imported energy during the 1970s reversed slightly in 1980. However, over the decade of the 1980s, it is expected [12,20] that a small increased dependence on imported energy will occur, followed by a decreasing dependence in the 1990s. Sources of supply projected until the year 2000 are shown in figure 1.7,[20] based on information available up to October 1979. Other projections are similar,[12,21] with the most notable differences being in the opinions of how nuclear energy, coal, shale, and solar energy will be developed.[16,19] A rapid change is expected in the mix of primary energies used to meet demand in the 1980s. The change in the fractional makeup of the primary energy supply projected in reference 16 is shown in figure 1.8. The

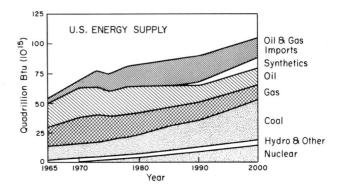

Figure 1.7 Historical and projected sources of primary energy used in the United States. (Data from reference 20)

12 Energy Use in the United States

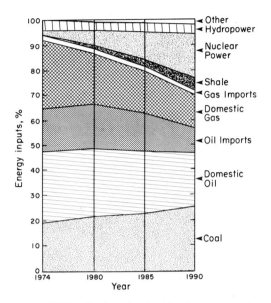

Figure 1.8 Distribution of projected primary energy inputs in the United States. (Data from reference 16)

validity of the prediction depends a great deal upon the development of nuclear power. Governmental predictions for installed nuclear capacity for generating electricity in 1990 have gone from 500 to 400 to 220 to 130 GW (gigawatts) in studies made in 1973, 1975, 1977, and 1979, respectively.[22] Development of nuclear power has not progressed as rapidly as technically and economically feasible, perhaps because of basic psychological fears and public doubts that it is a controllable technology.

Development of new raw energy resources for normal use is a long arduous process. Public concern about airborne effluents and land preservation affects the development of coal and shale resources. The requirements for satisfying publicly imposed regulations have increased the time delays involved in developing raw energy sources. New technological developments are often necessary to allow satisfactory development in a changing set of environmental restrictions. The economics associated with a given development project are affected considerably by these time delays. In 1972, it was believed that if the price of crude oil reached $8–10/barrel, shale oil could be developed as a replacement; 3500 quadrillion Btu (about 50 times the 1972 energy use) were considered identified and recoverable.[23] Even though the price of crude oil today is $36–41/barrel, precious little shale is being developed. Problems associated with technological development, water necessary in the shale-to-oil conversion processes, as well as changing environmental and economic considerations that indicate oil from shale will now cost about $40/barrel,[24] have militated against shale development. Even though the United States may be rich in energy resources, such as uranium, shale, and coal, shortages of forms of energy desired by end users are likely over the next few decades.

Estimates of US energy resources are made periodically by the US Geological Survey. Accuracy of the estimates are from 20 to 50% for identified recoverable resources. For undiscovered, submarginal resources, the estimates may be inaccurate by a factor of 10. Estimates are affected considerably by assumptions of market price and technological developments. As price increases, estimates of recoverable resources increase because it becomes economically feasible to extract low-grade or hard-to-recover sources of energy from the earth. Less expensive and more efficient recovery processes also tend to increase the quantities considered recoverable. Coal is the largest recoverable energy reserve in the United States, as depicted in figure 1.9.[25] More than 80% of the potentially usable primary energy resources in the United States is coal. This is the world's largest economically recoverable coal reserve—200–260 billion tons (about 6000×10^{15} Btu).[26] This is about 80 times the total energy consumed in the United States in 1980. About 815 million tons of coal were produced in 1980. In the 1920s, more than 70% of the US energy needs were met by coal. However, the cleaner and less expensive oil and gas resources displaced it from many applications. Over the past decade, new methods of handling and burning coal and flue gas purification have made coal use easier, and the prices of oil and gas have increased enormously. In March 1980 a utility company could buy $1.30 worth of coal to provide the same quantity of heat as $2.04 worth of gas or $4.30 worth of fuel oil.[26] As depicted in figures 1.7 and 1.8, coal is expected to contribute increasingly to the supply of US energy over the next 20 yr. It will be used directly to generate heat as well

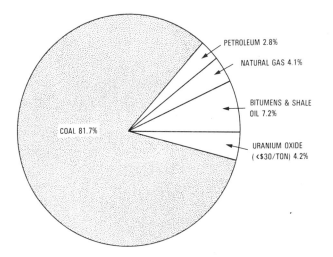

Figure 1.9 Distribution among presently known recoverable primary energy sources in the United States (Printed with permission of Dames and Moore from reference 25)

as indirectly to provide synthetic oil and gas. The United States is and will remain a major exporter of coal.

The petroleum resources in the United States are estimated at 800×10^{15} Btu (about 23 times the oil consumed in the United States in 1980).[24] Natural gas resources amount to about 400×10^{15} Btu. Uranium-based nuclear energy resources could yield about 200×10^{15} Btu. Shale could supply about 1000×10^{15} Btu to perhaps 10 times that figure at higher oil price levels. Hydroelectric power presently could be expanded only slightly. Tar sands have a potential yield of about 400×10^{15} Btu. Comparisons among quantities normally expressed as tons, barrels, cubic feet, and so forth, must be made by appropriate conversion to a universal energy unit. The British thermal unit (Btu) is the unit most commonly used. Not all oil, gas, coal, or uranium-bearing ore has the same Btu content. The conversion factors generally accepted[27] to represent the average Btu content of resources are

crude oil: 1 barrel (42 gallons) = 5,800,000 Btu
coal: 1 ton = 26,000,000 Btu
natural gas: 1 cubic foot = 1070 Btu
electricity: 1 kWh (kilowatt-hour) = 3412 Btu

All the above estimates of energy reserves are highly dependent upon assumptions about, among other things, price levels, technical developments, energy distribution networks, regulations, and public acceptance. Any estimate is subject to a great deal of uncertainty. Hence, one should use such figures only for relative comparisons between resources. It is wise also to make relative comparisons between values all presented from the same authority. Such a comparison is depicted in figure 1.10.[24] The vertical axis represents quantities of resources of any given type within the United States. The horizontal axis represents the price level at which a resource is considered recoverable in the quan-

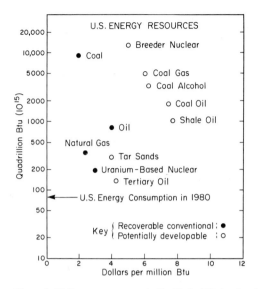

Figure 1.10 Energy resources in the United States developable at various energy prices. Quantities of a given resource increase as price increases. (Printed with permission of Dames and Moore from reference 24)

tities plotted. To add perspective, the beginning 1981 energy prices were approximately oil, $6.20/million Btu; natural gas, $2.60/million Btu; coal, $1.80/million Btu.

Less widely recognized energy sources that offer opportunity for future development include nuclear breeders, geothermal, wind, tidal, solar-thermal, photovoltaic, solid waste, biomass, and thermionic sources. More convenient fuels for the end user may be made from coal, agricultural crops, and shale. These include syngas, synoil, alcohol, and shale oil. Quantities that could potentially be developed at designated price levels are shown in figure 1.10. If the overall price levels exceed those plotted for each resource in the short term, the recoverable quantities should increase analogously to a normal supply curve as a function of price.

However, because the total quantity is ultimately fixed, the short-term supply curves would decrease year by year as the nonrenewable resources are depleted.

Renewable energy resources that arise from radiation and gravitational energy, reaching the earth from the rest of the universe, offer considerable potential for exploitation. In the very long term sense, many of the resources we commonly consider nonrenewable are actually renewable. Coal, oil, and shale can be considered concentrated forms of solar energy that has been gradually stored over the eons of time via plant and animal life cycles. Because of the increasing demands of humankind for naturally stored energy, it is being depleted faster than it is being replaced. Because the natural energy storage process is slow and relatively inefficient, it is advantageous to bypass it and make direct use of the solar and gravitational forces. Photosynthesis followed by burning or conversion to burnable alcohol, methane, or hydrogen is a shorter process. Direct use of wind, water convection, tidal waves, falling water, and temperature gradients are all more direct uses of solar and gravitational forces. Solar energy conversion directly to heat, from heat to electricity, or directly to electricity by the photovoltaic effect is a very direct use of a "renewable" energy source.

Because a renewable energy resource is not depleted, the quantity available for use cannot be depicted in the manner shown in figure 1.10. The quantities thought to be developable annually under the estimated price considerations are illustrated in figure 1.11.[24] Limitations on energy availability from renewable resources come from, among other things, restrictions on surface suitability, the technological development rate, price relative to other sources, and natural interruptions such as cloud cover. From a comparison of figures 1.10 and 1.11, it is obvious that the price of conventional fuels will need to rise even more before the alternative

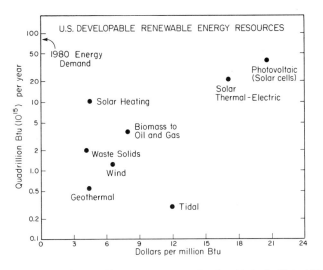

Figure 1.11 Renewable energy resources developable in the United States at various energy prices. (Printed with permission of Dames and Moore from reference 24)

resources become cost competitive. On the other hand, technological progress expected in the next two decades should reduce the cost of usable energy from solar radiation to a very competitive position.[28,29]

Secondary Energy: Electricity

Secondary energies are those forms of energy that have gone through a conversion from a primary fuel to a form more suitable for use by the end user. Synthetic fuels may be considered secondary energy supplies. Electricity is the most used secondary energy source in the United States. This section is devoted entirely to the consideration of the supply of and demand for electricity. Because electricity is the form of energy supplied by photovoltaic energy converters (solar cells), the demand for electricity and alternative ways of supplying it are important to the prospects for solar energy conversion by photovoltaics.

Nearly all electricity used in the United States is supplied by large utility companies. Exceptions to this occur where portability or remote use is required. Primary batteries (those used once and discarded), such as the carbon-zinc Le-Clanche "flashlight" and alkaline-manganese, mercury, and lithium cells, are used for powering calculators, flashlights, cameras, toys, radios, clocks, and so forth. Secondary batteries (rechargeable), such as lead-acid and nickel-cadmium batteries, are used for starting, lighting, and ignition in automobiles, boats, and aircraft, and to power shavers, fork lift trucks, golf carts, electric vehicles, toys, and so forth. Secondary batteries are generally recharged by electrical power from electrical power utilities or from engine-powered generators. Besides utility- and battery-supplied electricity, other sources are generators driven by diesel or gasoline engines, particularly of the industrial size used in remote locations; wind-driven generators; small water-driven generators; and solar cells.

The present annual use of utility-provided electricity in the United States is about 2.3×10^{12} kWh.[30,31] For 1980, the average price was 4.5¢/kWh. Hence, this is about a $100 billion business. Use of electricity grew at more than a 7.5% annual rate from 1900 to 1970. During the 1970s, the rate of growth decreased but still exceeded the overall growth rate in energy use. The consumption of electricity in the United States since 1900 is illustrated in figure 1.12.[2] Incorporation of the lower growth rate during the 1970s yields a historical growth rate of about 7%/yr. If this were extended into the future, about 7.5×10^{12} kWh would be needed in the year 2000—triple the present consumption. Meeting

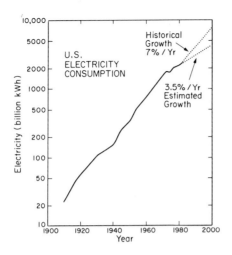

Figure 1.12 Historical and projected consumption of electricity in the United States. (Based chiefly on data from reference 2)

such a demand requires much long-term planning and confidence in projections because the lead times for constructing electrical power-generating plants are 6–8 yr for coal-fired and 8–12 yr for nuclear plants.

More than 60% of US electricity is used for commercial and industrial uses. Hence, much of the demand is caused by business activity. A good measure of business activity is the GNP. When the GNP in constant dollars is plotted versus electricity use, a linear relation is found with minor year-to-year deviations. This relation is shown in figure 1.13 for the years between 1950 and 1980. The slope of the graph is 3.2; that is, the use of electricity increased at 3.2 times the constant dollar GNP increase over the last three decades. If such a trend were to continue and the growth in GNP were 2% then a 6.4% annual increase in demand for electricity would be envisioned. Such an extrapolation is probably not warranted, however. Electricity as a secondary energy source

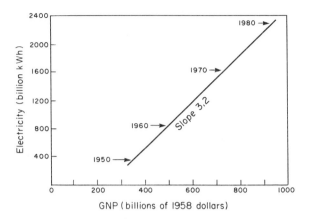

Figure 1.13 Historical relation between annual electricity use and the GNP in the United States. (Based chiefly on data from reference 2)

has been in its infancy during the twentieth century, and one would expect the growth rate in use to be highest during the industry's infancy. Availability of electricity has now been extended to all but the most remote areas of the United States.

On the other hand, there are several considerations that might induce present consumers of coal, oil, and gas to convert to electricity. It is clean and convenient at the site of the end user. It offers energy on demand by the flip of a switch. Also, for long-term reliability of supply, electricity offers a certain advantage because it can be generated by using any one of the primary fuels. If one fuel becomes scarce or extremely costly, utility companies are likely to convert to an alternative fuel, thus saving the consumer the expense and inconvenience of changing his own facilities. This tends to stabilize the price of energy to the consumer. As illustrated in figure 1.14, the energy flows in the US energy system afford great flexibility in choice of primary fuel to drive the consuming sectors—if electricity is used as a secondary en-

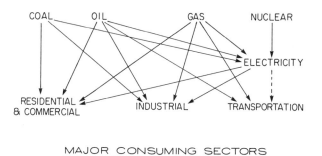

PRIMARY FUELS

COAL   OIL   GAS   NUCLEAR

ELECTRICITY

RESIDENTIAL
& COMMERCIAL   INDUSTRIAL   TRANSPORTATION

MAJOR CONSUMING SECTORS

Figure 1.14 Interfuel substitutability for electricity generation and supply of major energy-consuming sectors.

ergy source.[32] It is only in the transportation sector that the expense of electricity storage and the technological state of the art militate against substantial powering by electricity.

An interfuel substitutability study[33] has shown that the fuels used in various sections of the nation are a function of their price and availability. Because primary fuel prices change at various rates, the changing relative prices among fuels can be expected to alter the mix of fuels consumed.[34] The general price levels of primary fuels over the past two decades are shown in figure 1.15. For ease of comparison, the prices have been converted to dollars per million Btu by using the energy conversion factors described earlier in this chapter. Although the price levels among fuels seem to increase in similar patterns, the more easily handled and cleanest-burning fuels command the higher prices. Of course, artificial price controls by cartels of oil-producing nations, US regulations on gas and oil prices, and governmental controls on coal mining and effluents from burning coal can cause relations between the prices of primary fuels that are abnormal with respect to those expected under a completely free enterprise system. Under the conditions presently foreseen by Chase Econometrics,[35] US crude oil prices, averaged over the 1980s, will increase annually 16.7%; natural gas, 18.1%; and coal, 7.4%. These projections are illustrated in figure 1.15. One might deduce that there will be a growing price incentive for utilities to burn coal to produce electricity.

In 1978 about 12% of the electricity used in the United States was generated by hydroelectric plants. Of the thermally generated electricity, the mix of primary fuels used was as shown in figure 1.16.[30] Slightly over half was generated by burning coal. Even though about 70% of the generators could be powered by coal, various factors militate against full utilization of that capacity. Even though the relative prices of primary fuels may generally be as shown in figure 1.15, the levels at a given location may differ. Also, the necessary stack gas desulfurization required to meet ef-

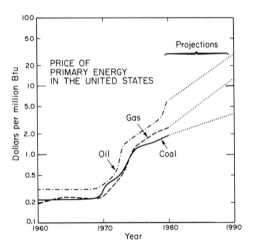

Figure 1.15 Historical and projected prices of primary fuels in the United States.

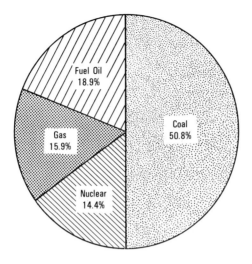

Figure 1.16 Mix of primary fuels used in thermoelectric generation in the United States during 1978. (Data from reference 30)

fluent standards may add too much economic burden to allow conversion from gas or oil to coal on a local plant basis. More than half of the power stations have multiple fuel capability.[36] In recent years, effluent regulations as well as local scarcity of specific fuels have resulted in increased conversion of single-fuel stations to multifueled plants. Because of the variations in fuel, labor, and capital cost factors, the price of electricity may differ by 400% across the United States. Typically, in states where a high proportion of the electricity is generated in hydroelectric plants, the price is relatively low. Unfortunately, there are few potentially developable hydroelectric power sites left in relation to the growing power needs.

The interrelations between factors that affect the electricity generation industry are complex. Various factors affect demand and tend to force increased utilization of one primary fuel relative to others. The main factors in the system can be schematically represented in a systems dynamics[37,38] model as shown in figure 1.17. In this diagram, the sign beside each arrow represents a decreasing ($-$) or an increasing ($+$) effect of the causing factor at the tail of the arrow on the affected factor at the point of the arrow. Taking the demand for electricity as the main point of interest, the diagram shows that population, use per capita, economic activity (GNP), price and reliability of electricity, use of electricity in environmental control, and competition from primary fuels in space heating directly affect the demand for electricity (arrows pointing toward it). In the absence of overriding factors, the greater the demand and utilization of electricity, the lower will be the primary fuel reserves. The lower the reserves, the greater will be the cost of fuel and greater the price of electricity, thus lowering the demand. This is a self-regulating negative feedback loop. Likewise, the lower the

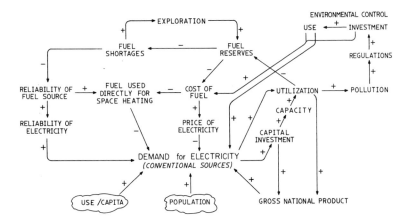

Figure 1.17 Schematic representation of the conventional thermoelectric generation system. An increase in the level of a cause (tail of arrow) results in an increase ($+$) or decrease ($-$) in the level of an effect (point of arrow).

primary fuel reserves, the greater will be the fuel short-ages—resulting in lower reliability of fuel and electricity. The lower the reliability of supply of electricity, the lower will be the demand for it. Increased utilization of electricity, particularly that generated by burning coal, leads to more pollution, which in turn yields more regulation to control it. This leads to greater investment in electrically powered stack gas cleaning equipment, which leads to increased demand and utilization of electricity. This is a positive feedback loop with no self-correcting mechanism. The latter loop can be kept from going out of control by various factors, including conserving, deregulating, and converting to a less polluting primary fuel, such as natural gas.

Such systems dynamics models could be diagramed for each primary fuel source and then integrated, yielding a model that would reflect the dynamics of crossover from one fuel to another as respective supplies and prices change. Other factors affecting such crossover would include the rate of progress in stack gas desulfurization technology, public confidence in nuclear power safety, development of low-sulfur coal resources, development of fuel transportation systems, and development of synthetic fuels. Increased reliability of supply, decreased cost, and decreased pollution associated with a particular fuel would increase the demand for it relative to other fuels. Because the cost of primary fuel has become a more and more significant portion of the cost of electricity over the last decade, future prices for electricity will depend more and more upon which primary fuels are used to generate it. Hence, there will be an increasing influence of the price of the primary fuel on the decision to use it to generate electricity. Figure 1.18[18] depicts the mix of primary energy resources used to generate electricity from 1965 to 1975 and a 1975 projection of the mix to be used from 1975 to 1990. In fact, nuclear and coal utilization up

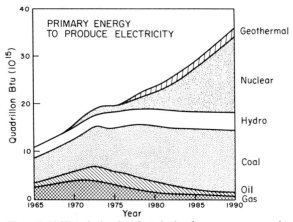

Figure 1.18 Historical and projected mix of energy sources used to produce electricity in the United States. (Data from reference 18)

to 1980 has not grown as much as predicted, with the difference being filled by oil and gas. The overall trends, however, are still valid predictions for the 1980s.

Although interfuel substitutability increases the reliability and price stability of electricity relative to a given primary energy source, electricity is still an expensive source of energy because its generation relies on a relatively inefficient use of coal, oil, and gas (35–40%) and, also, the capital costs involved in generating and distributing it are considerable. It is, however, the most economical method of using hydro- and nuclear power. During the development of the electrical utility industry, the price of electricity decreased considerably. The trend in price of electricity to residential customers as a function of time is shown in figure 1.19.[30] Also shown is the annual consumption of electricity per residential customer over the same time period, 1926–1980. Figures available in reference 31 indicate that for the year ending in June 1980, the average price per kilowatt-hour in residential use was 4.60¢ versus 4.14¢ the previous year

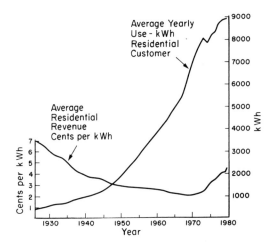

Figure 1.19 Historical residential use and price of electricity in the United States. (Data from references 30 and 31)

(increased by about the inflation rate). It is evident from figure 1.19 that demand increased considerably (6.6% yr), as the price of electricity was decreasing in the 1925–1970 period. However, as the price increased from 1970 to 1980, the demand increased at the much slower rate of about an average of 2.2%/yr. While this trend occurred in the residential sector (33% of the electricity market), similar trends occurred in the industrial and commercial sectors (41% and 22% of the electricity market, respectively). The average price of electricity in 1980 was only slightly greater than in 1970 in constant dollar terms (2.1¢/kWh versus 1.8¢/kWh in 1970 dollars). However, any change upward relative to the long steady decline in price of electricity is bound to affect the demand for it.

Taking all of the above into consideration, along with various prognostications of future primary energy costs, a reasonable guess of the average annual growth rate in electricity use is 3.5%/yr, as shown in figure 1.12. Similar values have been projected by others.[16,20] This is about double the projected growth in primary energy use. If this scenario is fulfilled over the next 20 years, the US electrical utility business in the year 2000 will amount to $170 billion in 1980 dollars. Capture of a small fraction of this business is a considerable incentive for entrepreneurs in the solar cell business.

References

1. E. Cook, *Sci. Amer.* 225 (1971): 137.

2. R. C. Dorf, *Energy, Resources, and Policy* (Reading, MA: Addison-Wesley, 1978).

3. Joint Committee on Atomic Energy, *Certain Background Information for Consideration When Evaluating the National Energy Dilemma* (Washington, DC: US Printing Office, 1973).

4. National Petroleum Council, *U.S. Energy Outlook, A Summary Report of the National Petroleum Council* (December 1972), p. 15.

5. National Petroleum Council, *U.S. Energy Outlook, A Report of the National Petroleum Council's Committee on U.S. Energy Outlook* (December 1972).

6. Shell Oil Company, *The National Energy Outlook* (March 1973).

7. US Department of the Interior, *United States Energy Through the Year 2000* (December 1972).

8. Associated Universities, Inc., *Reference Energy Systems and Resource Data for Use in the Assessment of Energy Technologies* (Report to US Office of Science and Technology, under Contract OST-30; Document AET-8, April 1972).

9. National Petroleum Council, *U.S. Energy Outlook, An Initial Appraisal 1971–1985* (July 1971).

10. J. A. Merrigan, *Sunlight to Electricity: Prospects for Solar Energy Conversion by Photovoltaics* (Cambridge, MA: MIT Press, 1975), pp. 1–13.

11. P. J. Bernstein, Newhous News Service, *Times Union* (Rochester, NY: 31 December 1980): 1A.

12. Gulf Oil Corporation, *Annual Report* (1979), pp. 4,5.

13. J. H. Krenz, *Energy: From Opulence to Sufficiency* (New York: Praeger, 1980), pp. 12,13.

14. N. McBride, *Mid American Outlook* (Cleveland, OH: CleveTrust Corporation, 1980), pp. 10–12.

15. Bureau of the Census, *Projections of the Population of the United States 1975 to 2000* (Washington, DC: US Department of Commerce, 1975).

16. Energy Group, *Capital Resources for Energy through the Year 1990* (New York: Bankers Trust Company, 1976).

17. J. Eibling, *Solar Energy: An Assessment for Business* (Columbus, OH: Battelle Memorial Institute, 1979), p. 4.

18. Exxon Company, U.S.A., *Energy Outlook 1976–1990* (December 1975).

19. Federal Energy Administration, *National Energy Outlook* (Washington, DC: US Government Printing Office, 1976).

20. Public Affairs Department, *World Energy Outlook* (New York: Exxon Company, 1980), pp. 8,9.

21. G. O. Barney, *The Global 2000 Report to the President* (Washington, DC: US Government Printing Office, vol. 1, 1980).

22. D. J. Lootens, "The Nuclear Option," *Engineering Bulletin* 51 (1980): 18.

23. P. Theobald, S. Schweinfurth, and D. Duncan, *Energy Resources of the United States* (Washington, DC: Geological Survey Circular 650, 1972).

24. P. Gotlieb, "Alternative Fuels," *Engineering Bulletin* 51 (1980): 34.

25. C. E. Mann and W. D. Bullock, "The Outlook for Coal," *Engineering Bulletin* 51 (1980): 11.

26. W. Glasgall, Associated Press, *Times Union* (Rochester, NY: 10 September 1980): 6D.

27. E. H. Thorndike, *Energy and Environment: A Primer for Scientists and Engineers* (Reading, MA: Addison-Wesley, 1976), p. 273.

28. ERDA-49, *National Solar Energy Research, Development, and Demonstration Program* (Washington, DC: US Government Printing Office, 1976).

29. W. E. Morrow, Jr., *Technology Review* 76(2) (1973): 31.

30. Edison Electric Institute, *Statistical Year Book of the Electric Utility Industry* (New York, 1978).

31. Edison Electric Institute, *Source and Disposition of Electricity* (New York: Edison Electric Institute, Volume 48, No. 6, 18 September 1980).

32. D. C. White, Energy Laboratory, *Final Report Submitted to the National Science Foundation—Dynamics of Energy Systems* (Cambridge, MA: MIT, 1973).

33. T. D. Duchesneau, Federal Trade Commission Economic Report, *Interfuel Substitutability in the Electric Sector of the U.S. Economy* (Washington, DC: US Government Printing Office, 1972).

34. J. Griffin, *Bell J. Econ. Mgmt.* 5 (1974): 515.

35. Business Bulletin, *Wall Street Journal* (New York, 4 September 1980): 1.

36. A Report of National Economic Research Associates to the Edison Electric Institute, *Fuels for the Electric Utility Industry 1971–1985* (New York: Edison Electric Institute, Publ. No. 72–27, 1972).

37. J. W. Forrester, *Principles of Systems* (Cambridge, MA: Wright-Allen Press, 2nd preliminary ed., 1972).

38. J. W. Forrester, *Industrial Dynamics* (Cambridge, MA: MIT Press, 1961).

2

Solar Energy as a Resource

Quantity of Sunlight

The sun has been and will continue to be a long-term source of energy for the earth, lasting billions of years. This huge power source is fueled by hydrogen, which undergoes fusion into helium with concomitant emission of radiant energy.[1] The temperature in the center of the sun is estimated at 30,000,000°F. The surface temperature is about 10,000°F. The mass of the sun is 2.2×10^{27} metric tons. About 1.4×10^{14} tons, or $6 \times 10^{-12}\%$, is consumed annually in the fusion reactions. The continuous radiation from the sun amounts to 3.8×10^{23} kW, or 3.6×10^{23} Btu/sec. Of this energy, less than one-half billionth reaches the earth's atmosphere. The solar energy that reaches the near earth atmosphere is 1.73×10^{14} kW, or 1.64×10^{14} Btu/sec. Annually this amounts to 5.1×10^{21} Btu.

The solar energy that reaches the atmosphere and the disposition of it are depicted in figure 2.1.[2] About 47% of it is absorbed by the earth and the atmosphere, thus generating heat. It is reemitted to space as longer-wavelength ir (infrared) radiation, resulting in cooling. About 30% of it is reflected from the earth and atmosphere back out into space. About 23% goes into evaporation and precipitation processes and ultimately is reemitted to space as long-wavelength radiation. Only 0.215% is used in causing waves, winds, and other thermally induced convection currents. Even less, 0.023%, goes into photosynthesis in plant growth. Looking at the earth as a single point in space, it is held in energetic equilibrium by radiant and gravitational energies flowing to and from it. The sum of the reflected solar energy, the reemitted infrared energy coming from substances heated by the sunlight and other sources—such as waste heat from burning, geothermal sources, and chemical and nuclear energy—is essentially equal to that solar and gravitational energy that interacts with the earth from space.

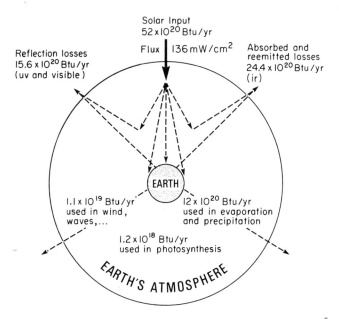

Figure 2.1 Schematic representation of solar energy input to the earth's atmosphere and disposition of it. (Based chiefly on representations in reference 2 and reference 2 of chapter 1)

The mean tidal power input associated with gravitational forces amounts to about 9.6×10^{16} Btu/yr, a small amount relative to the solar energy falling on the earth.

The solar energy that is used in photosynthesis, about 1.2×10^{18} Btu/yr, is responsible for energy-dense fossil fuels, such as oil, coal, gas, and shale, that we use so conveniently today. These fuels are produced in the long and inefficient natural processes of plant and animal growth and decay. However, wood and dry crop wastes have an energy content on a weight basis about two thirds that of coal. Raw biomass contains little sulfur or ash and, except for handling difficulties, can be easily burned, gasified, fermented to alcohol, or converted to methane for use as a renewable source of energy.[3] The key limitation to using these resources on a large scale is the low efficiency of plants as solar energy converters. Although there is a theoretical efficiency of 6.6% in photosynthesis, the typical efficencies are 0.3–3%, depending on the vegetation used.[4] Hence, rather large land areas that are suitable for plant growth would be required. Such use of the land would tend to conflict directly with normal use for food production.

The present annual world energy consumption is $2-3 \times 10^{17}$ Btu. This is a very small fraction of the solar energy impinging on the earth's atmosphere: 5.2×10^{21} Btu. The solar energy that penetrates the atmosphere and reaches the earth's surface in the United States is 5.13×10^{19} Btu/yr. This is 675 times the US consumption of energy in 1980. Solar energy is free and is distributed widely, if somewhat intermittently, across the nation. It is obvious that sunshine has the basic potential for satisfying much of the US energy demand. It is left to the ingenuity of mankind to discover or invent ways to use it economically in a more direct way than by the natural photosynthetic processes.

Terrestrial sunshine is a source of rather diffuse, low-temperature energy. Variations from day to night, day to day, and season to season make it necessary to have a means of energy storage or an alternative source of energy to use during dark periods. The solar power density* is 17 W/ft² (watts per square foot) in the continental United States averaged across the nation for a period of 1 yr.[5] The power density varies from 0 to a maximum of about 100 W/ft² at sea level on a surface perpendicular to the sun at noon on a clear day. This varies slightly with altitude, approaching 126 W/ft² at the upper edge of the atmosphere (solar constant = 136 mW/cm²). The amount of power that can be generated by a solar energy converter depends on the amount of sunlight it intercepts. The solar power available is found by multiplying the converter area by the solar flux or power density irradiating the converter. For example, in direct sunlight of 100 W/ft² power density, 10 kW of power is available in an area of 100 ft². The relation of various power uses to areas needed to intercept the equivalent in direct sunlight (100 W/ft²) and average sunlight (17 W/ft²) is illustrated in figure 2.2. Of course, the efficiencies of solar energy converters as well as many other factors must be considered when assessing how much converter area is needed for an application. It appears, however, that even though solar energy is rather diffuse, it can be converted with reasonable array areas into sufficient power for many uses.

Of course, the average solar incidence is higher in some

---

*Power is expressed in terms of Btu per second, joules per second (watts), kilowatts (kW), megawatts (MW), horsepower (hp), and so forth. Sunlight is described in terms of power density, the power in a given area, Btu per second per square foot, watts per square foot, milliwatts per square centimeter, and so forth. See the appendix for conversion units and definitions.

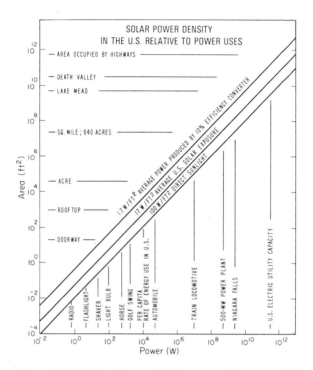

Figure 2.2 Relation of areas of solar energy collectors equivalent in capacity to readily recognized power uses and sources.

regions than in others. Figure 2.3[6] illustrates the distribution of yearly average and December average solar energy that falls on a horizontal surface in various regions of the contiguous 48 states. The southern and especially the southwestern portions of the United States, typically called the sunbelt, receive the most sunlight and hold most promise for cost effective solar energy utilization. Figure 2.4 illustrates the average monthly sunshine hours in selected cities in these regions.[7] The incident solar energy in the United States is least during December, January, and February. The solar radiation during December is about one half the average over

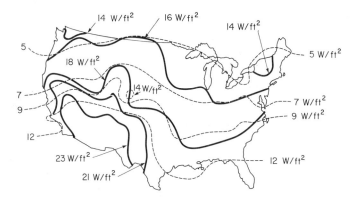

Figure 2.3 Distribution of solar energy over the continental United States: December average, - - -; yearly average, ———. (Data from reference 6)

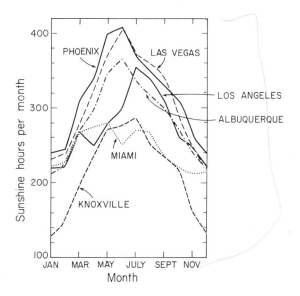

Figure 2.4 Average monthly direct sunshine hours in selected cities. (Data from reference 7)

a year. This seasonal variation can be ameliorated somewhat by proper orientation of solar-absorbing media. For example, depending upon the location and type of collector, about 30% more annual energy collection and 80% more winter collection can be achieved by orienting a previously horizontal collector to face the south at 45° above horizontal. Collectors that track the sun intercept about 80% more energy over a year and 150% more during the winter.[6]

To add perspective, if by tilting an absorber 45° to the south, we collected 1.3 times the average annual incidence of solar power on the US surface (17 W/ft²), we would collect an average of 22 W/ft². On a 1000-ft² collector, typical of residential roof areas, a total of 22,000 W would be collected. With conversion to electricity at 12% efficiency (typical of solar cells), a total of 22 kW $\times$ 0.12 $\times$ 8766 hr/yr, or 23,142 kWh, of electricity would be produced in a year. This is more than double the average yearly residential usage in 1979: 8901 kWh.[8] Without the benefit of electrical storage in batteries, flywheels, or such, the amperage in a 110-V (volt) household system would vary from 0 to 109 A (amperes) with an average of 24 A. Zero corresponds to nighttime. The maximum, 109 A, is calculated by multiplying 1000 ft² by the highest flux of sunlight when the sun is high in the sky and perpendicular to the collector (100 W/ft²), then multiplying by 0.12 in efficiency and dividing the total by 110 V. A typical household circuit allows for up to 150 A, with the average amperage over a year being 9.2 A (8901 kWh/yr $\div$ 8766 h/yr $\div$ 110 V).

At an average of 22 W/ft² collected sunlight and 12% conversion efficiency, the entire demand for electricity in the United States in 1979, 2.3×10^{12} kWh, could be generated by using an area of 10^{11} ft². This is an area 60 by 60 miles, that is, 3600 square miles. This is only 0.1% of the US land area (3,615,000 square miles). Using 12% as an average conversion efficiency, the entire annual energy usage of 76×10^{15} Btu could be provided by 10^{12} ft^2, an area of about 190 miles by 190 miles, or 36,000 square miles, 1% of the US land area. Currently, highways cover about 0.7% of the United States. Single-family residences account for 0.05%. Harvested croplands occupy 14%.[9] A very high percentage of the land is untillable and unsuitable for normal use. The sunshine falling on desert areas is greater than the national energy demand. An interesting fact is that the solar energy falling on Lake Mead, the lake formed by Hoover Dam, is five times the output from the hydroelectric generators at the dam.[1]

Quality of Sunshine

Solar energy is plentiful relative to the various uses of energy, but it is diffuse and must be collected and stored to make it readily available on demand in modern-day life. Typically, sunshine is converted to heat via absorption in materials. This can give rise to thermally induced work, by winds, waves, and convection or by man-made devices that capitalize on the heat. Sunlight can also cause photosynthesis, evaporation, and photoelectric currents in certain materials. In all cases, the spectral quality of sunlight relative to the spectral absorption and reflection characteristics of the absorbing medium is important. The basis for sunlight conversion into more useful energy is its absorption in properly selected media. Light can be considered to be either wavelike or corpuscular, depending upon whichever view works best in a particular case. When it is considered to be corpuscular, the smallest particle of light is a photon. Specific physical laws have to be obeyed in absorption of a photon by a material. For example, for absorption to occur, an excited energetic state equal to the photon energy must be allowed in the absorber. Conversion between photon energy and wavelength of light is made by using the equation wavelength (in μm) = 1.234 μm eV (electron volts) ÷ photon energy (in eV). To maximize solar absorption, the material characteristics should be matched to the spectral distribution of sunshine.

The extraterrestrial spectrum of sunlight in close to that of a blackbody radiator at 5800K.[10,11] However, because the atmosphere absorbs a disproportionate amount of short-wavelength light (high-energy photons), it is generally accepted that the shape of the solar spectrum at sea level most closely fits that of a 6000K radiator in the wavelength region below 1.25 μm.[12] The coarse sunlight spectra above the atmosphere (AM0) and at sea level (AM1) in the United States are shown in figure 2.5.[13] The predominant portion of the energy in sunlight lies in the visible region, 0.38–0.75 μm. At sea level with the sun overhead in a clear sky at normal humidity, the total radiant flux is reduced by 20–30% from that above the atmosphere, depending upon atmospheric conditions. Values of the energy distribution in the spectrum are typically

	<0.4 μm	0.4–0.8 μm	0.8–1.0 μm	>1.0 μm
Extraterrestrial %	9	49	13	29
Sea Level (AM1) %	6	57	14	23

Under these conditions, the energy falling on the atmosphere is about 126 W/ft² and at sea level about 100 W/ft². Considering silicon as an absorbing material with a band gap of 1.1 eV, one can calculate the energy that can be absorbed. If a nonreflective, but transmitting, coating over the surface reduces the reflection to 0, and if the thickness is such that all

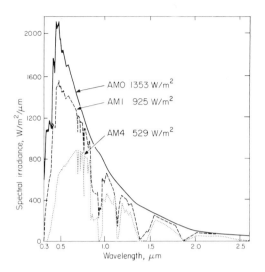

Figure 2.5 General spectral characteristics of extraterrestrial sunlight AM0 and sunlight after passing through one atmospheric thickness AM1 and four atmospheric thicknesses AM4. (Representation based on data in reference 13)

photons of energy greater than the band gap are absorbed, the solar energy absorbed can be approximated as follows. The wavelength in micrometers (μm) above which negligible absorption occurs (that is, photons having energy less than the band gap energy 1.1 eV) is 1.234 μm eV ÷ 1.1 eV, or 1.122 μm. Looking at the above table and figure 2.5, one can estimate that about 80% of the sunlight would be absorbed, that is, 0.8 × 100 W/ft^2, or 80 W/ft^2.

The radiant flux reaching an absorber is subject to the variations of weather and the angle between the surface of the absorber and the sun. At angles other than normal to the sun, the flux is reduced. Sunlight conditions are often designated in air mass (AM) numbers, where the numbers refer to the secant of the angle between the position of the sun at the zenith and at the position under consideration (measured at sea level). With the sun at the zenith, the condition is AM1. When the sun is lower in the sky by 60° from the zenith, the condition is called AM2. It may be easier to consider the air mass number as the number of atmospheres that the sunshine has penetrated to reach the point of measurement. AM0 refers to sunlight just outside the earth's atmosphere. Of course, reflections from clouds or particulate air pollution can reduce the intensity considerably. Absorption by typical variations in atmospheric gas mixtures does not influence the total energy reaching the ground to a great degree, but does alter the spectrum, particularly in the uv (ultraviolet) and ir (infrared) regions (figure 2.5). Ozone absorbs strongly from 0.2 to 0.35 μm and some at 0.6, 4.7, 9.6, and 14.2 μm. Carbon dioxide absorbs strongly at 2.7 and 4.3 μm and less from 13 to 17 μm. Water vapor is the chief barrier to atmospheric transparency in the infrared, absorbing strongly at 1.4, 1.85, 2.7, and 6.3 μm and from 14 μm to longer wavelengths. Besides these strong absorptions in the relatively less energy rich regions of the solar spectrum, well-resolved spectra[10,11] reveal small, narrow absorption bands in the visible region for water vapor and oxygen. These bands are 0.594, 0.652, and 0.723 μm for water and 0.629, 0.688, and 0.762 μm for oxygen. The coarse spectrum in the visible region, where most of the energy is concentrated, is nearly constant from location to location, and the direct, focusable radiation (about 60% of the total on average) has nearly the same spectrum as reflected sky radiation.

Standardization of solar reference conditions used to compare experimental results between researchers in the field of terrestrial photovoltaic energy conversion is very difficult. Various proposals have been made[13-15] with some consensus that an AM1.5 solar spectrum be taken for comparison under practical midday conditions. Typically, values in the literature for solar cell efficiencies refer to simulated AM1 conditions, which are provided in the laboratory by filtering through water the light from a tungsten-halogen lamp with 3200K color temperature and adjusting the intensity until a reference radiometer indicates a power density of 100 mW/cm^2.

Conversion of Sunlight to Other Energy Forms

Conversion of sunlight to other energy forms useful in doing work or providing warmth involves absorption of photons with concomitant transfer of the photon energy to the absorbing medium. This absorption of energy typically causes an increase in vibrational and electronic energy levels in the absorbing medium on a molecular or elemental basis. The larger the flux of sunlight and the more absorbing the material, the more will be the accumulated absorption of energy and the warmer the medium will become. This gives rise to variations in temperature at the surface of the earth that cause evaporation, wind, waves, and conductive and convective currents, which can be utilized in many ways to provide useful energy.

Neglecting natural uses of sunlight, such as for growing food and passive heating, the most commonly used method of converting sunlight to conveniently usable power is by absorbing it in black materials, which are thereby heated. This thermal energy is transferred to water or gas flowing through or by the black absorber. A well-designed heater has one or two glass or plastic windows over the absorber to reduce convection and infrared radiation losses. Of the 80–85% of the light transmitted through the windows, 80–95% is absorbed by the black absorber, with a resultant 65–80% collection of the incident energy. Losses inside the collector from convection, conduction, and reradiation bring net collection efficiencies down to 50–75%. With such efficiencies, water and space heating of homes can be practical in many parts of the United States.[16] Millions of small solar water heaters are used around the world for residential hot water supplies. About 10,930,000 ft^2 of solar heat collectors were sold in the United States in the year ending June 1978. About 60% went to heating swimming pools, 20% to water heating, 19% for space heating and cooling, and 1% elsewhere.[17] The main impediment to growth in this use is the initial cost of collectors. Collectors having daily efficiencies of 30–70% range in price from $20 to $100/ft^2. A simple swimming pool heater with about 20% efficiency sells for $4–$6/ft^2.

More sophisticated systems involving parabolic collectors and higher temperatures may compete with fossil fuel burning for providing heat and electricity to industrial parks and in base-load electricity for public utilities.[3] Generally, a 60% solar collection efficiency and a 40% efficiency for steam-powered turbines are assumed, with a resultant 24% overall efficiency for electricity production. Although this sounds like a reasonable efficiency, the cost associated with providing solar collectors as well as the associated steam generators put the projected cost of such facilities at over $2000/kW. This is higher than the $800–$1200/kW required for a fossil-fueled plant, but the latter uses rather expensive primary fuels that add to the operating cost. Solar thermoelectric generators are expected to compete favorably by the end of this century.

Conversion of solar energy to electricity in an economically competitive way is a desirable goal because of the ease of use of electricity without detrimental effects on the environment. Fortunately, in certain types of materials, photon absorption results in direct generation of electricity. These semiconductive materials have the required electronic band structure to allow direct separation of the photoexcited electrons and photoholes that are produced during photon absorption. This gives rise to a voltage gradient across the material. Thus, the material acts like a photoactivated battery that produces electrical power. The effect is referred to as the photovoltaic effect. Direct photovoltaic conversion of sunlight to electricity is an enticing prospect for clean production (using no moving parts) of electrical power. At present, the main limitations are low efficiency of conversion and high costs of solar cells (photovoltaic converters). Large- and small-scale terrestrial facilities can convert solar energy to electricity with an efficiency of about 12%. In cost (about $10,000/kW) and efficiency, terrestrial photovoltaic power stations presently do not compare favorably with solar-powered thermoelectric generation.

In both of the above cases, energy storage is necessary for use in nonsunlit hours. An alternative scheme[18] for providing large-scale supplies of electricity from solar energy is to build a photovoltaic station in synchronous orbit, where there is a nearly continuous supply of sunlight of about eight times the average intensity on the ground. Using microwaves, the transmission efficiency from direct current in space to direct current on earth[19] would be 55–75%. Then overall efficiency using cells with 15% efficiency in space might be in the 8–11% region before distribution of the electricity. This would appear to be a rather exotic way to solve the problem of intermittency effects in solar energy at the earth's surface. However, in the long term, availability of space vehicles and appropriate manufacturing technology might make it technically feasible to build such large power plants in space.[20]

There are many technical and economic considerations that relate to the utilization of photovoltaic energy conversion as a pollution-free means of meeting the energy demands of the United States. Technical progress is being made, demand for electricity is increasing, the costs of primary fuels are increasing at an unprecedented rate, and the costs of solar cells are decreasing rapidly. The rest of this study will deal with the present technological state of the art, the markets for photovoltaic conversion systems, and economic and business projections. It will be shown that even though sunlight presently is not being harnessed effectively as an energy source, technological progress and demand for alternative sources of energy will make it a well-utilized direct source of electrical power by the end of this century.

Wait

References

1. D. S. Halacy, *The Coming Age of Solar Energy* (New York: Avon Books, 1975), pp. 23–38.

2. S. S. Penner and L. Icerman, *Energy: Demands, Resources, Impact, Technology, and Policy* (Reading, MA: Addison-Wesley, vol. 1, 1974), pp. 144–146.

3. W. D. Metz and A. L. Hammond, *Solar Energy in America* (Washington, DC: American Association for the Advancement of Science, 1978), pp. 103–123.

4. ERDA-49, *National Solar Energy Research, Development, and Demonstration Program* (Washington, DC: US Government Printing Office, 1976).

5. NSF/NASA Solar Energy Panel, *An Assessment of Solar Energy as a National Energy Resource* (University of Maryland, December 1972).

6. MIT Energy Laboratory and MIT Lincoln Laboratory, *Proposal for Solar-Powered Total Energy Systems for Army Bases* (Massachusetts Institute of Technology, July 1973).

7. H. Landsberg, H. Lippmann, Kh. Paffen, and C. Troll, *World*

Maps of Climatology (New York: Springer-Verlag, 2nd ed., 1965).

8. Edison Electric Institute, *Statistical Year Book of the Electric Utility Industry* (New York, 1978).

9. R. S. Caputo, "Toward a Solar Civilization," *Solar Power Plants: Dark Horse in the Energy Stable*, R. H. Williams, ed., (Cambridge, MA: MIT Press, 1978), pp. 73–93.

10. R. M. Goody, *Atmospheric Radiation* (Oxford: Clarendon Press, 1964), pp. 417–426.

11. C. W. Allen, *Quart. J. Roy. Met. Soc.* 84 (1958): 307.

12. S. T. Henderson, *Daylight and Its Spectrum* (New York: American Elsevier, 1970).

13. A. P. Thomas and M. P. Thekaekara, *Proc. Joint Conf. Am. Sect. ISES and SES Can. Inc. Winnipeg* 1 (August 1976): 338.

14. K. W. Böer, *Proc. Joint Conf. Am. Sect. ISES and SES Can. Inc. Winnipeg* 1 (August 1976): 264.

15. *Intern. Comm. on Illumination CIE No. 20(TC–2.2)* (Paris: Bureau Central de la CIE, 1972).

16. K. W. Boer, *Chem. Eng. News* (29 January 1973): 12.

17. P. Sageev, D. Comini, and J. Mortland, *Solar Energy: An Assessment for Business* (Battelle Technical Inputs to Planning Review No. 2, 1979).

18. P. E. Glaser, *Space Resources to Benefit the Earth, Third Conference on Planetology and Space Mission Planning* (The New York Academy of Sciences, October 1970).

19. W. C. Brown, *Microwave Power Transmission in the Satellite Solar Power Station System* (Raytheon Company Technical Report ER72–4038, 1972).

20. B. Horovitz, *Industry Week* (26 May 1980): 68.

3

Principles of Photovoltaic Energy Conversion

Materials and Solid State Mechanisms

Most solar cells are composed of crystalline semiconductors with solid state characteristics that promote separation of photocarriers and a resultant flow of electricity in an attached external circuit. Although some are made from cadmium sulfide (CdS), cadmium telluride (CdTe), gallium arsenide (GaAs), germanium (Ge), amorphous silicon (SiH_x), and other combinations, the predominant basic material is silicon (Si). This is the substrate in most of the solar arrays used to power earth satellites. It is well defined theoretically and experimentally.

Each atom in a pure silicon crystal has four valence electrons that are shared with adjacent silicon atoms in covalent bonding. If the crystal is doped with an impurity such as phosphorus, which occupies the same lattice sites but has five valence electrons, the doped crystal contains valence electrons in excess of a pure crystal—one for each of the phosphorus atoms. At a given temperature, many of these excess electrons are separated from the phosphorus atoms by thermal energy and are free to wander in the crystal, making it an electron-conducting, n-type semiconductor. If silicon is doped analogously with boron, which has only three valence electrons, there will be one electron too few to complete the covalent bonding in the vicinity of each boron atom. This electron vacancy, or absence of an electron, appears to the lattice to be positively charged, because an electron would normally occupy that site. The electron vacancy, or hole, is a positive charge carrier that, when thermally detached from the boron impurity, is free to wander in the crystal, making it a hole-conducting, p-type semiconductor. Because positive nuclear charges balance the negative valence electron charges in both p- and n-type semiconductors, there is no macroscopic charge disequilibrium in or on the crystals.

If the p-type and n-type silicon crystals were figuratively joined together to make a perfect single crystal, the electrons in the n-type portion would diffuse across the joining boundary (p-n junction) into the "electron deficient" p-type region, and the holes would likewise diffuse into the "hole deficient" n-type region, until a voltage equal to the sum of the diffusion potentials of the holes and electrons was established across the p-n junction. Thus a permanent electrical field is established in the region of the junction. In practice, junctions like this are made by ion implanting, diffusing, or otherwise growing p-type impurities into an n-type crystal or n-type impurities into a p-type crystal.

The resultant electronic band structure of interest is shown schematically in figure 3.1. By virtue of the electron and hole flows described above, the conduction and valence bands in the p-type material have risen relative to those in the n-type region. Because the Fermi level, which represents the electrochemical potential,[1] was originally higher in the n-type crystal, the level is equalized within the "joined crystal" and across the p-n junction by virtue of the n-type material feeding electrons to the p-type material and vice versa with respect to holes. The p-type portion now contains a disproportionate amount of electrons and is at a more nega-

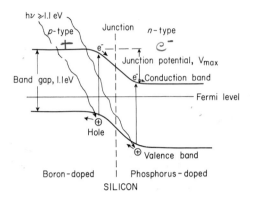

Figure 3.1 Electronic band structure at a p-n junction in silicon solar cells.

tive potential (higher in the figure), which causes an electrical field across the junction. If light of greater energy than the valence-to-conduction band gap falls upon the crystal, it may be absorbed by valence band electrons, which are thus excited to the conduction band, leaving a vacant electronic state, a hole, in the valence band. Under the influence of the electrical field, the photoexcited electrons will be driven toward a lower energy state in the n-type portion of the crystal, while the holes move toward a lower energy state, which for them is in the p-type region. The photocarriers have moved into respective regions of the crystal where like charges are the majority charge conductors, a photovoltage is created, and a current can flow in an external circuit; this is pictorially described in figure 3.2. This defines the photovoltaic effect. Materials designed to provide this are called solar cells, photovoltaic converters, photoelectric cells, or photocells.

Not all photons irradiating this specially prepared silicon crystal give rise to external current I_{load} in the load resistance R_{load}. There are various loss processes that reduce the effi-

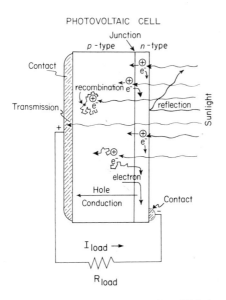

PHOTOVOLTAIC CELL

Figure 3.2 Schematic representation of light interactions and photoelectric current flow in a silicon solar cell.

ciency of conversion of sunlight to electricity. Out of the spectrum of photon energies that comprises sunlight, only photons with energy greater than the difference in energy between the conduction and valence energy bands, that is, the band gap (E_g; 1.1 eV for Si),[2] can be absorbed to produce photocharges. This amounts to about 80% of the incoming solar light in the case of Si (see chapter 2). Part of the light is reflected off the surface of the crystal because of the differing refractive indices of Si (3.4) and air (1). The reflectivity R is given by $(3.4 - 1)^2 \div (3.4 + 1)^2$, which is 30%. Thus, approximately 56% of the photons in sunlight are of the proper energy, are not reflected, and can enter the crystal to be absorbed. The fraction absorbed in the crystal is $1 - \exp(-\alpha th)$, where α is the absorption constant, *th* is the thickness of the crystal, and $\exp(-\alpha th)$ is the fraction transmitted through the cell. The optimal depth of the *p-n* junction beneath the face of the crystal is dictated by such factors as where in the crystal most of the light is absorbed so as to produce electron and hole pairs, the lifetimes and mobilities of the photocarriers, and the resistance of the very thin side of the junction next to the surface through which the current must flow to reach the front contact electrode.[3] This latter resistance is governed principally by the geometry of the metal grid contacting the surface of the crystal, whereas the other phenomena are a function of the material characteristics of the crystal itself. A defect-free single crystal is needed to enhance the lifetimes of the photocarriers, prevent their recombination (a loss process), and allow them to diffuse further. This increases their chances of encountering the electrical field at the junction, being separated, and producing a current. The absorption coefficient α varies as a function of wavelength. For red light of 0.7 μm wavelength, α in Si is about 2000 cm^{-1}, which, when substituted into the absorption equation, yields a thickness of 3.5 μm required to absorb one half of the incident photons. Typically, solar cells made from silicon are about 250 μm thick with the *p-n* junction 0.5 μm to a few micrometers below the front surface.

The collection efficiency is a measure of the proportion of minority carriers (holes in the *n*-type and electrons in the *p*-type regions) produced by absorbed photons that reach the junction. Of those carriers generated outside the influence of the electrical potential at the junction, some diffuse toward the junction while others diffuse away and recombine in the bulk of the crystal or at the surface. In typical silicon solar cells, the collection efficiency ranges from 60 to 80%.

Another loss process involves conducting electricity through the very thin resistance layer between the *p-n* junction and the front side electrical contact. The geometrical factors of contact location, say, using a grid or screen instead of a spot contact, and junction depth, which partly governs the resistance, have to be balanced against loss of incident light through masking and collection efficiency deterioration. Typically, the efficiency of the cell is reduced several percentage points by this series resistance.[4]

The voltage that can be developed by a solar cell is a function of the excess of minority carriers on each side of the p-n junction. This voltage is less than the band gap because of junction losses. The voltage developed increases with intensity of illumination toward a condition in which the minority carrier density approaches the majority carrier density, and the voltage approaches the band gap energy. The voltage could never go beyond that because the junction potential would be nullified. In reality, it never approaches the band gap energy very closely because, as soon as the junction potential (figure 3.1) is counterbalanced by separated photocarriers, no internal field exists, and maximum voltage or open circuit voltage, corresponding to an infinite value of R_{load} (figure 3.2), is reached. This voltage is directly proportional to the band gap energy. The junction loss (about 0.5 V in Si; 45%) decreases exponentially with increasing band gap. The loss also decreases as the temperature decreases because of a decrease in thermally generated carriers that may reach the junction. The maximum voltage of an experimental silicon cell under full sunlight at room temperature is about 0.6 V. If the load resistance is designed to yield maximum power output (current times voltage), the voltage is about 0.45 V. Thus, even after photon absorption, the cell converts a relatively small portion of the absorbed energy (>1.1 eV per photon, the band gap) to useful energy. Theoretical limits on conversion efficiency involve other factors also and will be considered later in this chapter.

Material Absorptance

The absorption properties of photovoltaic materials considerably influence how much of the incoming solar energy is converted to electricity. This is true not only because of the fraction of the solar energy that is absorbed, but also because of where in the semiconductor the majority of photons are absorbed with respect to the surface and the p-n junction and how much energy is converted to heat instead of electricity. Referring to figure 3.1, only those photons with energy greater than the band gap are strongly absorbed. For silicon, this means that solar photons of greater energy than 1.1 eV (wavelengths less than 1.122 μm) are absorbed in a relatively short distance into the material. Photons of longer wavelengths are not absorbed readily, and hence the solar photon spectrum (figure 2.5) at wavelengths greater than 1.122 μm is not appreciably intercepted by silicon. Of the spectrum that is absorbed, 1.1 eV per photon goes to form a hole-electron pair, while energy in excess of 1.1 eV per photon goes to heating the crystal. For example, the photons of 0.561-μm wavelength have an energy of 1.234 μm eV ÷ 0.561 μm, or 2.2 eV, just double the energy needed to produce a photocarrier pair. Hence, 50% of the energy in the spectrum at 0.561 μm will go into heating and not into production of electricity.

The absorption spectra for Si and several other semiconductors used in photovoltaic devices[5] are shown in figure 3.3. The absorption edge corresponds to the band gap energy: for Si, 1.1 μm corresponds to a band gap of 1.12 eV; and for CdS, 0.52 μm corresponds to a band gap of 2.37 eV. Using the equation fraction absorbed = $1 - \exp(-\alpha d)$, where α is the absorption constant (in cm^{-1}) and d is the distance into the crystal (in cm), one can calculate that for 0.8-μm photons, 50% of the incident light is absorbed within

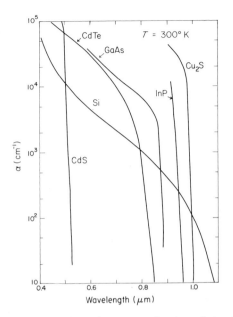

Figure 3.3 Absorption spectra of various photovoltaic materials. (Printed with permission of *Acta Electronica* from reference 5)

7 μm of the surface of Si and CdTe, and within 1 μm for GaAs, at 300K. The higher the absorption constant, the closer to the surface the absorption occurs; hence the *p-n* junction should be located accordingly. Surface effects and resistances are more critical in materials with higher absorption constants.

Comparison of figures 2.5 and 3.3 indicates that sufficiently thick semiconductors of lower band gap energy absorb a larger fraction of solar energy. The total flux of extraterrestrial photons and the fraction of incident photons having energy greater than the band gap decrease as the band gap increases, as shown in figure 3.4.[6,7] As the band gap decreases (absorption goes to longer wavelengths), the frac-

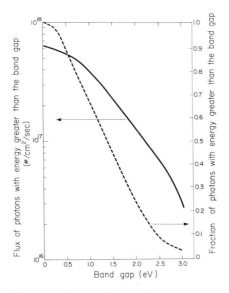

Figure 3.4 Flux and fraction of the photons in AM0 sunlight that can be absorbed in a thick semiconductor as a function of the semiconductor band gap energy. (Data from references 6 and 7)

tion of sunlight absorbed increases. Likewise the heat energy losses, due to absorption of photons with energy in excess of the band gap, increase as the band gap decreases. Thus on the basis of spectral considerations alone, there is an optimal band gap, for which the sum of the gain from increased solar energy absorption and the loss from heating is maximized.

Although reflection of sunlight from surfaces must be considered in the design of solar cells, it will not be treated here. It is merely recognized that thin, usually evaporated, coatings can be used to make losses from reflection negligible.

Efficiency

The photovoltaic cell can be visualized as an equivalent electrical circuit,[8] diagramed in figure 3.5. The resistances R,

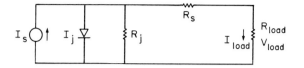

Figure 3.5 Simplified equivalent circuit diagram for a solar cell.

currents I, and voltage V correspond to the following in this figure: I_s = total separated charge; I_j = current leaked across the p-n junction, I_{load} = current in the load; R_j = junction resistance to leakage; R_s = series resistance of the cell, mainly from the thin surface layer; R_{load} = resistance of the load; and V_{load} = voltage across the load and hence the cell.

The maximum voltage attainable, V_{max}, is the same as the open circuit voltage of the cell, that is, when R_{load} is infinite. This is always a fraction of the band gap voltage at room temperature. The maximum power of an ideal solar cell is $I_s V_{max}$. V_{max} increases as a function of increasing band gap, while the number of photons in the solar spectrum capable of producing holes and electrons, hence I_s, decreases. Then the theoretical power maximum, the product $I_s V_{max}$, exhibits a maximum at a particular band gap energy.

Although I_s is fairly constant with respect to temperature (except for minor effects of band gap shifts), the junction resistance to leakage decreases, and current leaked, I_j, increases as temperature increases. Hence, the maximum voltage possible across the cell decreases with increasing temperature, and the power output, $I_{load} V_{load}$, decreases relative to the input of solar energy. Because of the complex manner in which V_{max} varies with temperature, the band gap at which the efficiency is optimal shifts to higher values as temperature increases.[9]

Figure 3.6 is an illustration of the variation of theoretical efficiency of solar cells made from semiconductors of in-

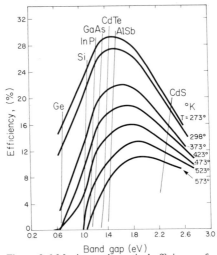

Figure 3.6 Maximum theoretical efficiency of solar cells as a function of band gap energy and temperature.

creasing band gap energy. It is typical of various theoretical treatments that agree in general but not in complete detail.[7-11] The efficiencies illustrated do not take into consideration any losses from cell resistance, photocarrier collection inefficiencies, reflections, transmission, or nonideal junctions. It is evident that the temperature at which a photovoltaic device operates considerably affects the efficiency of energy conversion and that the optimal band gap for maximum efficiency shifts upward with increasing temperature. The slanted, nearly vertical lines represent the shifts in band gap for a given semiconductor as a function of temperature. At room temperature, the maximum predicted efficiency is about 28% for a material of band gap about 1.4–1.5 eV, such as CdTe. The efficiency for CdTe is about halved when the temperature is raised to 200°C (473K).

The maximum power (mp) output of a solar cell is $V_{mp}I_{mp}$, which is usually somewhat less than the theoretical maximum, $I_s V_{max}$. As illustrated in figure 3.7, V_{max} and the current across an experimental silicon solar cell increase as light intensity increases.[10] At a given light intensity, by varying the load resistance one can derive a voltage-versus-current curve[12] from which V_{mp} and I_{mp} may be found. The variations of output voltage and photocurrent as the load resistance is varied on a typical silicon solar cell are plotted in figure 3.8. The current density is plotted in terms of milliamperes per square centimeter of the cell. For exposure to 100 mW/cm², I_s is 28 mA/cm² with zero load resistance. The maximum voltage is 0.6 V when the load resistance is infinite (open circuit). Hence, I_s multiplied by V_{max} yields a theoretical power of 16.8 mW for a 1-cm² cell. In practical use, when the load resistance is neither zero nor infinite, the resistance should be adjusted to yield a maximum in the product IV. Using figure 3.8 as the basis, figure 3.9 can be

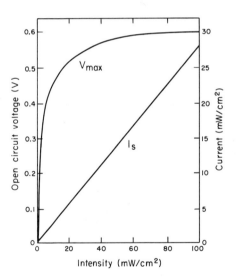

Figure 3.7 Voltage and current output of a typical silicon solar cell as a function of illumination. (Data from reference 10)

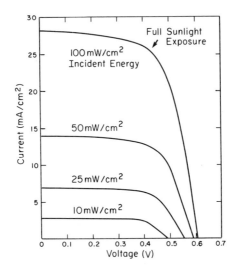

Figure 3.8 Variation in photocurrent and output voltage as the load resistance varies from zero (shorted) to infinity (open circuit).

derived to show that the maximum power under 100 mW/cm^2 illumination is 11 mW when V_{mp} is 0.46 V and I_{mp} is 24 mA. The maximum in power, $I_{mp}V_{mp}$, occurs at 0.46–0.40 V as the illumination decreases by a factor of ten. I_{mp} and V_{mp} always occur at the "knee" of the current-voltage curve (figure 3.8). The ratio of $I_{mp}V_{mp}$ to I_sV_{max} is called the fill factor. The fill factor is 11 mW ÷ 16.8 mW, or 0.65 in this case.

From figures 3.7–3.9, it is evident that the voltage generated by a silicon solar cell varies little over the normal range of daylight illumination conditions. This makes it ideal for charging batteries. The current, however, is directly proportional to illumination, and because the power is the product of voltage times current, the power output is proportional to the illumination. It is interesting that at low load resis-

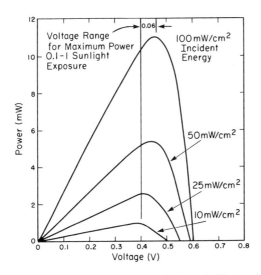

Figure 3.9 Power output of a typical 1-cm² silicon solar cell as the load resistance, hence the output voltage, is varied from small to large.

tances or when the cell is shorted, it acts as a constant current source and is not destroyed.

Power and efficiency are functions of the variation of V_{max} or V_{mp} and I_{mp} with temperature. These are illustrated in figure 3.10.[9] The percentage decrease of V_{max} with increasing temperature is greater than that for I_{mp}; hence, the decrease in efficiency with increasing temperature is attributable chiefly to the voltage variation. Figure 3.11 illustrates current-voltage curves for typical silicon solar cells under direct sunlight as temperature varies. The decrease in V_{mp} at the knee of the curves as temperature increases from 29 to 70°C is the main cause of a decreased power efficiency. The maximum power attainable under constant illumination at various temperatures relative to that at 25°C is plotted in figure 3.12.[13] The photocurrent does not vary much with temperature of the cell from −75 to 175°C.

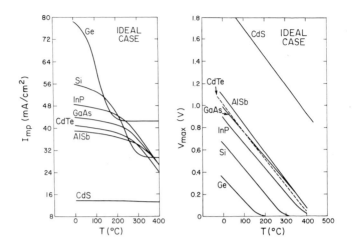

Figure 3.10 Theoretical variation of voltage and photocurrent outputs from selected photovoltaic cell materials as a function of temperature. (Data from reference 9)

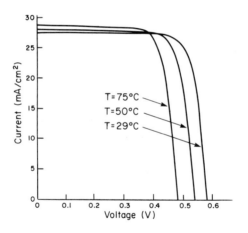

Figure 3.11 Characteristic photocurrent-voltage curves as a function of temperature for typical silicon solar cells under direct sunlight.

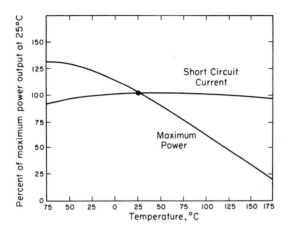

Figure 3.12 Power output and maximum photocurrent from typical silicon solar cells as a function of temperature relative to that at 25°C. (Data from reference 13)

The temperature effect on efficiency is less for semiconductors of higher band gap. Also, the higher the band gap, the smaller is the fraction of absorbed photon energy that goes to heating the semiconductor. In Si, this is more than 40% of the absorbed energy. Although Si may be more efficient at 25°C, CdS may be a better theoretical choice for a solar cell operating at 100°C or above. Of course, in practice the choice depends upon the state of the art in solar cell manufacture. In any case, one would prefer to keep the temperature of the photovoltaic cell as low as practicable. For large arrays, this may mean suitable engineering for heat transfer away from the cells by conduction and convection. Cooling the cells to below ambient temperature conditions is not energetically thrifty. Keeping the temperature near ambient may not be too difficult in the very diffuse energy flux from the sun. Cooling becomes more critical when focusing devices are used to concentrate sunlight onto a smaller solar cell. Concentrator cells specially designed to withstand concentration ratios of up to 200 are commercially available.[14] Because of the high intensity of sunlight impinging on them, the operating temperature of the cells is much higher than for normal cells. Heat transfer from the cells to water or air, which can be used in turn for heating, can add to the overall efficiency of solar energy conversion.

In summary, the efficiency of photovoltaic energy conversion with simple solar cells at ambient temperatures is not expected to ever exceed about 28%.

References

1. J. Tauc, *Photo and Thermoelectric Effects in Semiconductors* (New York: Pergamon Press, 1972), p. 18.

2. M. Wolf, "Limitations and Possibilities for Improvements of Photovoltaic Solar Energy Converters," *Proc. IRE* 48 (1960): 1246.

3. J. F. Elliott, "Photovoltaic Energy Conversion," *Direct Energy Conversion*, G. W. Sutton, ed. (New York: McGraw-Hill, 1966), pp. 1–37.

4. P. Rappaport and J. Wysocki, "The Photovoltaic Effect," *Photoelectronic Materials and Devices*, S. Larach, ed. (New York: Van Nostrand, 1965), pp. 239–275.

5. P. Rappaport and J. Wysocki, *Acta Electronica* 5 (1961): 364.

6. J. Wysocki, *Solar Energy* 6 (1962): 104.

7. R. Halsted, *J. Appl. Phys.* 28 (1957): 1131.

8. J. Loferski, *J. Appl. Phys.* 27 (1956): 777.

9. J. Wysocki and P. Rappaport, *J. Appl. Phys.* 31 (1960): 571.

10. P. Rappaport, *RCA Rev.* 20 (1959): 373.

11. W. Shockley and H. Wueisser, *J. Appl. Phys.* 32 (1961): 510.

12. R. Gold, "Current Status of GaAs Solar Cells," Transcript of Photovoltaic Specialists Conference, Vol. 1, *Photovoltaic Materials, Devices and Radiation Damage Effects* (DDC No. AD412819, July 1963).

13. Sensor Technology, Inc., *Data Sheet Number 185* (Chatsworth, California, 1980).

14. E. Robertson, ed. *The Solarex Guide to Solar Electricity* (Washington, DC: Solarex Corporation, 1979), pp. 50–52.

4

State of the Art in Photovoltaic Conversion Technology

There is no dearth of problems to be overcome before photovoltaic energy conversion competes economically with other means of supplying electricity on a broad basis. Technological problems include the proper control of manufacturing processes to yield reproducible, defect-free cells with high junction potentials, high collection efficiencies, and electrode contacts with low resistance. Engineering refinement is needed for building large, efficient solar cell arrays with the many electrical contacts needed and that will withstand climatic changes without degradation. And most important, methods for fabricating solar cell raw materials of adequate quality at a much lower cost must be found. If these problems are solved, another technological question will have to be answered before photovoltaic energy conversion can reach its highest potential. How can the electrical energy produced in the photovoltaic process be stored in an energy dense, conveniently redistributable way with high efficiency? Most electricity is used on demand, which does not always coincide with sunshine. Considerable scientific effort will be required to develop forms of energy storage, such as flywheels, fuel cells, or batteries that are more energy rich and less expensive.

There are, however, many reasons for optimism that the technological problems involved in making solar cells and arrays of them and efficient electricity storage can be solved. There has been much progress in increasing the efficiency of photovoltaic conversion since the modern solar cell with the *p-n* junction described in chapter 3 was invented in 1954.[1] The efficiency of silicon solar cells has increased from 6% to 15–19% over the last 25 years. Most of the power used in operating extraterrestrial satellites is provided by solar cell arrays that convert the continuous solar radiation in space to electricity for driving instrumentation. Besides the remarkable progress in development of silicon solar cells, over the last 5–10 yr a multitude of novel approaches to making photovoltaic energy converters have been developed. These have been based on various materials, including heterojunctions, CdS/Cu_2S, $(Cd/Zn)S/Cu_2S$, $CdTe/Cu_2S$, $CdS/CuInSe_2$, polycrystalline silicon, amorphous silicon, GaAs, Se, Schottky barriers of many types, and chlorophyll. Basically all such devices depend upon the appropriate solid state materials engineering to yield a permanent electrical field within the material in which photogenerated holes and electrons are separated to produce an external voltage across the material. The flow of the light-induced charge through an external load produces useful work in the same manner as produced by a battery. Materials purity and uniformity are important in all cases to reduce the probability of hole and electron recombination. Hence, special precautions are used in preparing raw materials and giving the cell the proper configuration.

The state of the art in solar cell development has been reviewed extensively by several authors.[2–10] The technological status in the main thrusts of solar cell development will be covered in this chapter, with attention to elements of cost and potential for technological and economic improvements.

Silicon-Based Homojunction Solar Cells

Photovoltaic cells produced by appropriate doping of homogeneous semiconductors to yield a *p-n* junction within the semiconductor are called homojunction cells. The common silicon solar cell made from a single crystal wafer is an example. Such a cell is not strictly homogeneous because the impurity dopant concentration is different as the *p-n* junction is crossed. However, because the main structure of the entire cell is silicon, the present convention is to call it a homojunction cell.

Practically all the solar cells used in spacecraft as well as in terrestrial uses are made from silicon. Silicon is the second most abundant element in the earth's crust (as SiO_2) and is produced in metallurgical grade for about $1/kg (kilogram).[11] Basically, quartzite is mined and reduced to silicon by carbon in submerged electrode arc furnaces. The molten silicon is taken from the furnaces and treated with oxygen, or a mixture of oxygen and chlorine, to help purify it to a metallurgical grade. After solidification and pulverizing the metallurgical-grade silicon, the fine particles are treated in a fluidized bed of hydrogen chloride and a copper-containing catalyst, which converts it to $SiHCl_3$. Trichlorosilane is then decomposed on a heated silicon surface to form polycrystalline semiconductor-grade silicon, often referred to as polysilicon. The world production of silicon is 500×10^6 kg/yr.[12] It is the critical substrate used in solid state electronics. The metallurgical grade, 99% pure, is not typically used in integrated circuits or in solar cells. The more refined and perfect semiconductor grade, 99.999% pure, is required for the best devices. It sells for $65–70/kg.[13] This type of silicon is grown in ingots 2–6 in. in diameter that are drawn slowly out of a hot melt of pure silicon. Wafers of silicon 150–500 μm thick are delicately cut from the ingot to avoid formation of defects by mechanical abrasion. These wafers are subjected to etching, electron bombardment, or other techniques to remove surface defects. They are next heated in the presence of an appropriate chemical that allows diffusion of a dopant into the surface to yield a *p-n* junction. The wafer may be subjected to more etching to adjust the surface-to-junction distance. Next, electrodes are attached, often by elaborate and tedious procedures. This is followed by coating with antireflection coatings, for example, SiO_x, SiN_x, TiO_2, ZnS, and CrO, and protective encapsulation. The cells are connected in series and parallel to meet the photovoltaic array output design characteristics, and the array is ready for use. This description is much too simplistic to convey the actual art and intricacy of the processes used. For high-efficiency cells, very elaborate crystal growing and regrowing procedures are used. Complicated profile doping by ion implantation and several vacuum metallizations to make electrodes are often used.

It is easily deduced that making solar cells in the above way is an expensive process. Even before the wafers of silicon are cut from the ingot, the cost of the silicon is $65–70/kg. If the wafers are 200 μm thick, a square foot contains $3 worth of silicon. If the waste involved in etching, sawing, mistakes, and so forth amounts to 50%, then it takes $6 worth of silicon to make the square foot of silicon cells. Currently a 1-ft^2 array of silicon solar cells, which yields 10% solar conversion efficiency, sells for about $100. This needs to be reduced to $2–10 before broad use of these types of cells will occur.

In considering whether such cost reduction is feasible, we can break the costs into components associated with raw materials, the process of growing the silicon, the fabrication process, and the protective and supporting structures. Of course, the higher the efficiency, the less will be the area needed to provide a given electrical output. Efficiency of 10% or more is needed for most practical applications. As discussed, the cost of semiconductor-grade silicon is about $70/kg. There are various technical approaches that indicate that the cost of solar cell-grade silicon can be reduced in the next few years to $1–10/kg.[12–14] Recently it was demonstrated that metallurgical-grade silicon, which costs about $1/kg, could yield solar cells with reasonably high efficiency.[15] This would provide a 60-fold reduction in cost of the basic starting material. The material in the silicon wafers 200 μm thick by 1 ft^2 mentioned would cost about 5¢.

The process of growing silicon and slicing it into wafers is at present very labor intensive and wasteful of material. Typically, 3-in.-diameter ingots of silicon are sliced into perhaps 25 wafers per inch of ingot. Sixteen or more would typically be mounted into a 1-ft^2 array and interconnected appropriately. This leaves about 25% of the surface bare of solar-converting cell material. For highly efficient arrays, the wafers are trimmed into rectangles to fit together in a closely packed array to provide more active area per square foot. Silicon "bricks" recently have been cast and subsequently cut into rectangular wafers to yield semicrystalline silicon wafers that yielded high-efficiency cells.[16] Also, a square die has been used as a shaping device while growing silicon ingots.[17] By creating isotherms of the appropriate shape and seeding with a (100) plane face silicon crystal, an ingot approaching a square shape can be drawn from molten silicon at 1410°C. A heat exchanger method has been developed that yields cubes of silicon more than 1 ft^2 in cross section.[18] This process is reported to save one third of the power, labor, and materials used in the more commonly used Czochralski round-ingot-pulling process.[15] Sophisticated ingot-slicing equipment also is being developed to cut 100 wafers per inch of ingot versus the 25 commonly used. Thus, a 400% increase in efficiency of materials utilization is expected from this technique alone.

The new methods for making cells from metallurgical-grade silicon, growing 1-ft^2 ingots, and cutting four times more wafers per inch of the ingots should yield more than a 300-fold decrease in cost of wafers. Besides this, the larger wafers should decrease dramatically the labor involved in making arrays of cells. The reliability of the arrays should be greater because fewer cell-to-cell electrical contacts would be required.

After inherently p- or n-type silicon wafers are prepared, they are typically etched to provide a defect-free surface and then baked in the presence of PH_3 or B_2H_6 gases, respectively, to allow diffusion of the dopant into the silicon to form a p-n junction. This can be done by high-energy ion implantation and annealing as well. Careful control of this step is required to yield high-efficiency cells.

The cells are next electrically contacted front and back with special care and coated with an antireflection coating. The integrity of each cell and of each cell contact is especially important to the efficiency of solar arrays because in interconnected cells the random variation in characteristics causes the maximum power from the array to be lower than the sum of the maximum powers that can be obtained from each individual cell.[19,20] The connecting grid on the face of a cell should be designed for power maximization resulting from the highest electrical conductivity at the least grid shadowing of the solar cell. If metallization contact is not adequate, there may be excessive contact resistance to the silicon and insufficient coverage to control photocharge losses in the front surface of the cell.[21,22] The cost associated with reliable electrical contacting and protecting of cells is a substantial fraction of the cost of an array.

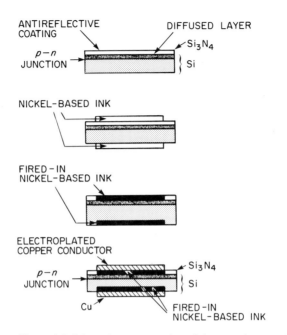

ANTIREFLECTIVE
COATING

DIFFUSED LAYER

Si₃N₄

p–n
JUNCTION

Si

NICKEL–BASED INK

FIRED–IN
NICKEL–BASED INK

ELECTROPLATED
COPPER CONDUCTOR

p–n
JUNCTION

Si₃N₄

Si

Cu

FIRED–IN
NICKEL–BASED INK

Figure 4.1 Schematic representation of the steps in a novel method to contact electrically silicon solar cells. [Reprinted from *Electronics* (11 September 1980). Copyright © 1980. McGraw-Hill, Inc. All rights reserved]

A novel method for cost reduction and making the contacting process more reliable has been developed recently. It is depicted in figure 4.1.[23] The first step is to evaporate silicon nitride onto the cell to reduce reflection losses. Then a nickel-based thick ink is silk-screened onto the cell to form contacts in a grid pattern. The composite is heated in air, driving the nickel through the silicon nitride to form an ohmic contact to the silicon. Finally, a copper layer is electroplated over the nickel to provide low-cost, high-conductivity solar cell contacts. Techniques such as this, used in conjunction with large, square, metallurgical-grade silicon, may allow more than a tenfold reduction in the cost of silicon solar cell arrays with above 10% efficiency.

Many other new processes for preparing silicon substrates for use in mass producing solar cells are being investigated. Techniques have been found to grow ribbons or sheets of large-crystal silicon by rolling,[24] coating sheets and recrystallizing through heated or molten zones,[25] dendritic growth,[26] and edge-defined film growth.[11] An excellent review of these techniques was made recently.[27] Silicon cells of high efficiency need not be more than 100–150 μm thick;[28] hence, there is strong incentive to avoid the slicing of ingots and concomitant materials losses by preparing thin layers in the silicon-growing process. The principal problem encountered in most processes for making thin layers is that many small crystals of silicon are formed as opposed to a near-perfect single crystal. The various crystal boundaries in polycrystalline silicon lead to inefficiencies in the photovoltaic process, and high-efficiency cells are difficult to obtain.

Silicon ribbons are perhaps the closest to large-area single crystals in performance. Solar cells made from silicon ribbons typically range from 8 to 12% in efficiency.[29] Ribbons can be grown by a variety of techniques using dies that shape the ribbon crystal as it forms. Dies can be wetted or not wetted when in contact with molten silicon. For the wetted die case, the technique is called edge-defined film-fed growth (EFG)[30] or may be called capillary action shaping technique (CAST).[31] For the nonwetted die, the process is generally called the Stepanov technique.[32] The process is depicted in figure 4.2. Dies can be made from high-purity carbon in the wetted case, or quartz in the nonwetted case. These may need mechanical support to prevent deformation at molten silicon temperatures.

Ribbons also can be pulled from molten silicon by the web dendritic method.[33] Such materials have produced 15.5%

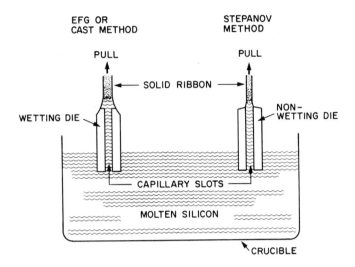

Figure 4.2 Schematic diagram of silicon ribbon-growing processes involving wetting and nonwetting dies.

conversion efficiency, with 1-ft^2 modules yielding up to 14%. A specially prepared seed crystal is used to achieve proper growth of silicon filament dendrites. Two of these filaments act as shaping edge guides for a ribbon pulled gradually from the surface of molten silicon. Ribbons several centimeters wide, 125–250 μm thick, and more than 10 m long have been achieved.[34] Because no dies or shapers are required, the silicon ribbon remains free of contamination. The uniformity of crystal growth requires careful control of growing conditions. This control can be facilitated by a gas jet shaping technique.[35] In the normal Czochralski ingot-pulling process, surface tension produces cylindrical ingots. This can be modified by the force of high-velocity gas applied to the silicon column as shown in figure 4.3.[35] By

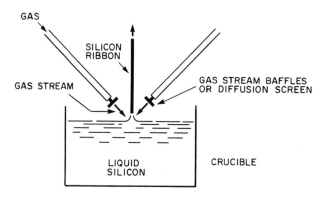

GAS

SILICON
RIBBON

GAS STREAM

GAS STREAM BAFFLES
OR DIFFUSION SCREEN

LIQUID
SILICON

CRUCIBLE

Figure 4.3 Schematic representation of a gas jet-shaping technique used to assist thin ribbon growing from molten silicon. [Reprinted from *Electronics* (6 November 1980). Copyright © 1980. McGraw-Hill, Inc. All rights reserved]

arranging gas jets so as to flatten the ingot into a ribbon, a thin crystal can be drawn from the crucible. Varying the gas temperature, pressure, angle, and number of jets can yield an equilibrium condition that allows formation of silicon ribbons of a desired width and thickness. All of the above silicon ribbon-growing techniques offer faster and more materials efficient means for silicon substrate preparation than by the ingot growing and slicing process. However, careful control of growing conditions is required, and the resultant rate of ribbon growth is slow. Typically, only a few inches per minute of the required quality ribbon can be grown.

Other means for preparing silicon substrates include dip coating, coating on a web, hot rolling, and chemical-vapor deposition.[36-39] Coating or deposition is generally done on a ceramic substrate such as $3Al_2O_3 \cdot 2SiO_2$. In chemical-vapor deposition, SiH_4 or chlorosilane compounds are thermally decomposed on the substrate in a low-pressure atmosphere. In a "floating substrate" technique,[40] silicon is chemically deposited on molten tin. A supersaturated silicon-tin solution provides silicon crystallites on the liquid tin surface. An equilibrium condition can be obtained whereby a crystalline silicon layer can be floated slowly off the surface, with new material being deposited along the receding edge of the film. In all of these cases, the silicon is produced in small, sub-micrometer crystallites. When solar cells are made from this polycrystalline silicon, the photovoltaic conversion efficiencies are considerably lower than those from single-crystal or large-crystal silicon. The photocurrent available in such cells is relatively low, indicating that recombinative photocarrier loss processes occur at the crystallite surfaces and interfaces. Because of the surface energetics involved, dopant concentration profiles needed to yield reliable p-n junctions are very difficult to control. A major improvement in efficiency can be attained either by increasing the crystallite size or by passivating the grain boundary states to reduce the number of photocarrier recombination centers.[41] The photocarriers that reach the p-n junction in a given crystallite are separated to yield photocurrent, but those that reach grain boundaries are lost. Experimental data indicate that there are about 1.6×10^{13} recombination centers per square centimeter of surface area on polycrystalline silicon.[42] This number may vary with preparation method or environmental history of the sample. The density of such sites should decrease as the grain surface-to-volume ratio increases and as the sites are passivated. Hence, larger crystallites and perhaps treatment with hydrogen, growth of protective oxide films, or other as yet undetermined means to passivate the recombination centers are needed to improve the performance of polycrystalline silicon solar cells. The increase in short circuit photocurrent and maximum photovoltage of such cells under sunlight has been predicted with a resultant cell efficiency dependence, as shown in figure 4.4.[41] Typically, polycrystalline silicon solar cells have less than 10% conversion efficiency.

There is considerable optimism that by understanding the loss processes in thin polycrystalline silicon cells, methods for producing high-efficiency cells can be developed. Laser annealing can be used to heal crystalline defects in silicon.[43] With a high-energy, short pulse of light by a laser of the appropriate energy, the surface micrometer or so of silicon crystals can be melted for a brief time.[44-46] This can be used to increase the size of crystallites. An increase from 0.02 to 1.5 μm has been reported in one case.[47] By using ion implantation techniques for preparing *p-n* junctions in polycrystalline silicon, one can achieve a more uniform doping than by the thermal diffusion processes. In the latter processes, the dopant materials appear to diffuse into the grain boundaries to yield channels of doped material extending

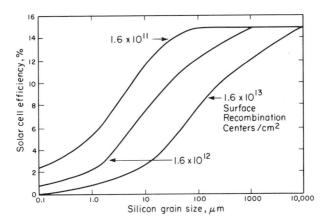

Figure 4.4 Predicted polycrystalline silicon solar cell efficiency as a function of silicon grain size at various concentrations of surface recombination centers. (Printed with permission of Elsevier Sequoia S.A. from reference 41)

through the substrate in some cases. This may cause current leakage, a low series resistance, and decreased efficiency. Ion implantation tends to yield doped material only at the horizontal surface. However, if normal heating is used to anneal the radiation damage caused by ion implantation, the dopants may still diffuse along grain boundaries. This problem can be avoided by annealing with laser pulses that melt the crystal for such a brief period that the dopants have too little time to diffuse significant distances. Hence, there are several approaches that can be taken to improve the photovoltaic conversion efficiency of these types of solar cells.

Because of the lower costs envisioned for mass producing polycrystalline films of silicon relative to wafers or ribbons, emphasis is being placed on this field of study. It is generally accepted that at least 10% conversion efficiency is required for photovoltaic converters to be practical for general use. Hence, increasing the efficiency of these materials is of primary concern. Significant progress has been made over the past 5 yr, with typically reported efficiencies increasing from 1 or 2% in the early 1970s to 10–12% in 1980.[48] The efficiencies of silicon solar cells typically reported in the literature as state-of-the-art cells over the past few years are illustrated in figure 4.5. This figure illustrates that although large-crystal cells certainly are more efficient, the state of the art in polycrystalline cells is progressing at a remarkably high rate and may be expected to be 10–15% in the 1980s. One should expect that commercial preparation of solar cells would yield average conversion efficiencies of terrestrial sunlight of about 70–80% of those values shown in figure 4.5. Commercially available arrays of cells, electrically connected in series and in parallel and protected from the environment, are typically 50–70% as efficient, as shown in figure 4.5. This is caused by, among other things, the normal

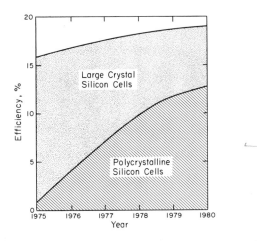

Figure 4.5 State-of-the-art photovoltaic conversion efficiencies reported for silicon solar cells from 1975 to 1980.

statistical variations in materials, process control, and connection losses.[19,20]

The foregoing discussion of homojunction solar cells centered around the silicon semiconductor. Other substances, such as gallium arsenide, gallium aluminum arsenide, and gallium indium arsenide, are used in homojunction configurations as well. Indeed, efficiencies of more than 20% have been achieved in these cells. Gallium arsenide has a band gap of 1.4 eV, nearly ideal for a single-cell configuration, and has a much higher optical absorption than silicon, thus enabling the use of thinner layers of semiconductor. However, high fabrication costs and scarcity of raw material relative to silicon have limited the use of these cells. Hence, they are intended mainly for use in solar concentration schemes for which large areas of sunlight are focused onto a small-area photovoltaic converter. These concentration cells will be discussed later in this chapter.

Heterojunction Solar Cells

In the previous discussion, a homojunction solar cell was described with the electrical field to separate photocarriers provided by the *p-n* junction within a silicon semiconductor substrate (see figures 3.1 and 3.2). Silicon and other semiconductors such as cadmium sulfide and gallium arsenide may be used for homojunction cells. An electrical field within a solar cell can be created in other ways. When two semiconductive materials are joined, at the interface between the two there is a mismatch in electronic band structure that, after electronic equilibration, yields an electrical field within the composite that can separate photogenerated holes and electrons. This is a heterojunction. Also, between a conductor (metal) and a semiconductor, an electronic equilibration of Fermi levels occurs that results in an electronic band "bending" at the interface, with an electrical field produced just below the surface of the semiconductor. This is a Schottky barrier. Hence, by vacuum deposition of metal onto silicon or epitaxial growth of one semiconductor onto another, an electrical junction can be formed similar in function to the *p-n* junction in the homojunction solar cells. A variation of the Schottky barrier is the metal-insulator-semiconductor (MIS) junction, which incorporates a very thin layer of insulating oxide between the conductor and the semiconductor. An insulating layer less than 20 Å thick can enhance the voltage output of a cell without lowering the photocurrent.[49,50] A very clear review of the formation of such junctions is presented in chapter 3 of reference 51.

A pictorial representation of the energy bands within semiconductors at junctions with other materials is shown in figure 4.6. The actual energy band profiles at interfaces of materials are very difficult to measure, but barrier heights

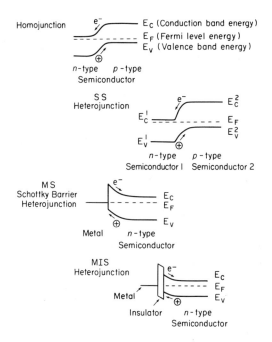

Figure 4.6 Schematic representation of electronic band structure in common heterojunction solar cells.

can be ascertained experimentally[52] and energy band profiles can be deduced. The junction energy band profiles at heterojunctions between two semiconductors depend upon the inherent energy band structure of each semiconductor, whether it is p- or n-type, and the relative positions of the energy bands. The latter can be derived by referencing to vacuum through the value of the electron affinity. The electron affinity is the energy gained by taking an electron in vacuum and placing it at the lowest level of the conduction band of the semiconductor. In the SS heterojunction illustrated in figure 4.6, semiconductor 1 is n-type (the Fermi level is closest to the conduction band), and semiconductor

2 is p-type (the Fermi level is closest to the valence band). Photoelectrons generated by light absorption in the electrical field region of semiconductor 2 near the junction move toward lower conduction band energy levels across the junction in semiconductor 1. Photoholes in the valence band move toward lower energies for holes that lie to the right of the junction in semiconductor 2. Thus the photocarriers migrate into regions where their charge is the majority carrier—electrons in the n-type semiconductor and holes in the p-type semiconductor. In this case, where the electrical field is present chiefly in semiconductor 2, one would expect the spectral response of the cell to be controlled mainly by light absorption in semiconductor 2. This can be affected considerably, however, by the relative mobilities and lifetimes of photocarriers in the two semiconductors. The nature of the junction potential profile, the spectral response, and the solar conversion efficiency are highly dependent upon the selection of semiconductors and preparation of the junction. With appropriate control of materials, the conversion efficiencies of these single-junction cells should be similar to those in homojunctions (see figure 3.6).

In the Schottky barrier junction, the magnitude of the energy barrier at the interface between the metal and the semiconductor is the difference between the electron affinity of the semiconductor and the work function of the metal. The work function is the energy required to take an electron from the Fermi level of the metal into vacuum. In the case shown in figure 4.6, the work function of the metal exceeds the electron affinity of the n-type semiconductor. Hence, after electrical equilibration of the Fermi levels between the two materials, a band "bending", as schematically diagramed, occurs within the semiconductor. Light absorption in the semiconductor will generate photocarriers that may be separated by the field just under the surface of the semiconductor, thus yielding the photovoltaic effect.

Schottky barriers can be formed by using p- or n-type semiconductors with a variety of metals. The energy band structure will vary accordingly. Typically, the metallic layer is vacuum deposited onto a thin semiconductor substrate in such a thin layer that it is highly transparent. Hence, light can pass through the metal and be absorbed in the semiconductor at the junction where the field exists. Alternatively, the semiconductor layer can be very thin, and light can be absorbed throughout its thickness with photocarrier separation near the back surface, where the metal is deposited. Solar cells made in this manner have about the same maximum theoretical solar conversion efficiency as homojunction cells,[53] about 22–24% with a silicon substrate and 25% with semiconductors having a band gap $E_C - E_V$ of 1.4–1.6 eV. The relative simplicity in fabrication of these cells makes them attractive. Of course, the normal electrical contacts of metal conductors to the surfaces of solar cells can give rise to Schottky barriers that may impede current flow across the cell. Hence, contacting grid and backing material must be selected carefully in combination with the semiconductor to ensure ohmic front and back surface contacts that do not interfere with current flow.

The formation of intimate metal-to-semiconductor electrical contact that is stable upon aging is an ongoing technical problem in making Schottky barrier heterojunction solar cells. The metal is typically vacuum deposited onto the semiconductor substrate in a 20–100-Å layer that allows transparency. A thicker metallic grid is evaporated onto the delicately thin layer to provide good electrical conductivity, and a protective, transparent, antireflection layer such as silicon nitride is evaporated over that. The metal-to-semiconductor contact often deteriorates because of exposure to gases in the environment. Oxidation at the surface of the semiconductor may occur by diffusion of oxygen through the thin metal layer or discontinuities in it. Performance can be improved if a uniform 5–20-Å-thick insulating layer is placed between the metal and the semiconductor. Basically, the thin insulating layer lowers dark current across the junction and enhances the voltage output without affecting the photocurrent.[4] Thicker insulator layers can cause deterioration of the maximum photocurrent attainable.[50,54] These modified Schottky barrier junctions are called MIS junctions: metal-insulator-semiconductor. The schematic energy band structure is presented in figure 4.6. The insulator is typically a layer of oxide such as SiO_2 on a silicon substrate. Electronic components based on similar materials and preparation processes, called MOS devices, are used commercially in computer memories with good reliability.

Solar cells based on the MIS junction have reached 17.6% efficiency with single-crystal silicon as the semiconductor.[55] More than 9% efficiency has been obtained with a thin film polycrystalline silicon substrate.[56,57] The cross section of a typical MIS cell is shown in figure 4.7. The variations possible in the design of heterojunctions have led to a proliferation of acronyms, such as SIS for semiconductor-insulator-semiconductor and CIS for conductor-insulator-semiconductor (covering MIS and SIS). In the lower section of figure 4.7, a relatively conductive semiconductor oxide has been used in place of the thin metal. The junction is formed across the thin oxide by the difference in work functions of the conductive oxide and the silicon. The conductive oxide is typically indium tin oxide (ITO), tin oxide, or cadmium tin oxide. The work function within the conductive oxide depends a great deal upon the method of preparation.[55] These layers can be formed by sputtering,[56,57] electron beam deposition,[58] and spray coating.[59] These types of cells are

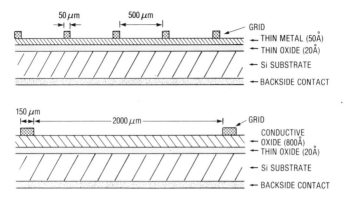

Figure 4.7 Typical layer structure in MIS (top) and SIS photovoltaic cells.

usually called SIS cells. Recent reviews of MIS and SIS cells[55,60] with silicon as the semiconductor indicate that the state of the art in these structures provides about 12–17% solar conversion efficiency with large-crystal silicon and about 9–12% with polycrystalline silicon substrates.[59–67] This is near to the state of the art using *p-n* homojunctions as shown in figure 4.5.

Multijunction cells of a variety of types have been designed particularly for use in converters that use light-concentrating mechanisms.[68–72] Schematic energy band profiles for n^+-n-p-p^+ homojunction cells and for graded-band heterojunction cells are shown in figure 4.8. In the first case, an increased efficiency relative to a single *p-n* junction cell might be achieved because electrical fields to separate photocarriers extend throughout a larger volume of the cell. In the second case, by choosing semiconductors of decreasing band gap with appropriate electron affinities and conduction properties as the cell is fabricated from semiconductor 1 to semiconductor 4, an extended electrical field is achieved. Because of the differing band gaps and light absorption, this is sometimes called a multicolor cell. Additionally, if illumination is through semiconductor 1, a more effective utilization of light may be achieved with more energetic blue light absorption near the front surface and less energetic red light absorption nearer to the back surface. Such a cell might yield a higher fraction of incoming light conversion to electricity and a lower fraction converted to heat relative to a single-junction cell. Cells such as GaAs/Si/Ge have been proposed.[73] Various analyses[74-76] indicate that such cells may have a theoretical efficiency of up to 40%, depending upon the choice of materials and the number of junctions used. Instead of considering semiconductor-to-semiconductor contact as in figure 4.8, we may consider simply overlapping several *p-n* junction cells. For economical manufacture, we might build these cells in a multilayer with the front semitransparent contact of one being the back contact to the cell above it, as suggested in the schematic dia-

MULTIJUNCTION CELLS

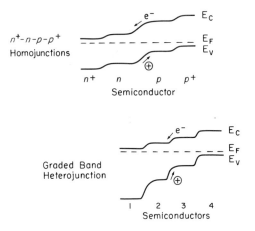

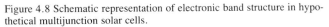

Figure 4.8 Schematic representation of electronic band structure in hypothetical multijunction solar cells.

gram in figure 4.9.[76] Such cells are called integrated tandem solar cells and act as though they were single cells connected in series. In a properly balanced two-cell integrated construction with a band gap of 1.1 eV for the back cell and 1.68 eV for the front cell, a 33% efficiency is theoretically possible. A three-layer construction with 1.1-, 1.46-, and 1.97-eV band gap semiconductors from back to front, respectively, yields a maximum theoretical efficiency of 37.6%.[76] In the limiting case, where there is a continuum of energy gaps across an infinitely absorbing cell, the efficiency of AM1 light conversion approaches 44%.[77] Of course, such devices would be difficult to make without resistance losses from the various contacts between materials. Relatively little experimental research has been devoted to the development of such complicated designs. However, one might expect increased sophistication of multijunction design with the potential for solar energy conversion efficiencies up to 40% in the future.[2] Other types of multijunction devices particularly designed for use under high levels of illumination will be illustrated in the concentration cell section of this chapter.

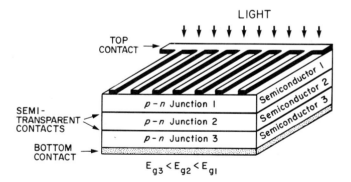

Figure 4.9 Integrated multilayer series solar cell structure connecting three cells with band gaps decreasing from front to back. (Printed with permission of Pergamon Press from reference 76)

Amorphous Silicon Cells

Amorphous silicon of the type used in solar cells was discovered in 1969 by decomposing silane gas (SiH_4) in a glow discharge.[78] It was found later that this "amorphous silicon" was actually an amorphous silicon-hydrogen alloy, which is currently referred to as hydrogenated amorphous silicon: α-Si:H.[79,80] It is typically deposited on a substrate by causing a plasma discharge between an anode and a cathodic substrate within an enclosed chamber containing 0.01–2.0 Torr of SiH_4. The plasma can be formed with radiofrequency or direct current sources. The α-Si:H can be formed also by sputtering high-purity silicon onto a grounded substrate in the presence of hydrogen at 3–20 vol % in argon at a total pressure of 0.005–0.010 Torr.[79,81] In the unhydrogenerated form, α-Si contains many unsatisfied chemical bonds that cause a high density of electronic energy states between the conduction and valence bands. These act as recombination centers; hence, the material has very low photoconductivity. Also, introduction of p- or n-type dopants has little effect on such material. However, the hydrogenated amorphous silicon shows high photoconductivity and responds to attempts to move the Fermi level by doping. For example, n-type α-Si:H can be produced by adding PH_3, AsH_3, $Sb(CH_3)_3$, or $Bi(CH_3)_3$ to the SiH_4 discharge or by ion implantation of P, As, Sb, Bi, Na, K, Rb, or Cs; and p-type α-Si:H can be produced by adding B_2H_6 to the SiH_4 discharge or by ion implantation with B, Al, Ga, In, or Tl.[82-84] Undoped material is typically slightly n-type as prepared. The conductivity can be varied from about 10^{-12} Ω^{-1} cm^{-1} (Ω is units in ohms) for intrinsic α-Si:H to 10^{-2} Ω^{-1} cm^{-1} by adding about 1 vol % of B_2H_6 or PH_3 in the silane discharge atmosphere.[82] Hydrogenated amorphous silicon typically contains 5–50 atomic percent hydrogen[85,86] with the formation of predominantly Si-H bonds, thus helping to satisfy the stoichiometry of chemical bonding in α-Si. Electron diffraction measurements indicate that α-Si:H has a short-range ordering of Si atoms similar to that in crystalline silicon, but longer-range (second-nearest-neighbor) distances between atoms are less than in crystalline silicon and are random.[87,88]

Because of the rather random nature of the atomic ordering in α-Si:H, it has a very high absorption coefficient. Depending upon the conditions under which it is deposited by glow discharge, it may have absorption in the visible spectrum of 10–20 times that of crystalline silicon,[80] and it may have a band gap of about 1.58–1.8 eV. Doping can lower the band gap significantly. In some cases involving sputtering and evaporative deposition, narrower band gaps are obtained. Both the strong light absorption and the nearly ideal band gap energy make α-Si:H appealing for making solar cells. It offers the possibility of very thin cells of perhaps 1–5 μm thick compared to 100 μm or so of crystalline silicon required to absorb sunshine adequately. Also, because it is amorphous, α-Si:H is free of the grain boundaries that interfere so much with the efficiency of polycrystalline silicon cells. And last, it is deposited by a relatively inexpensive process.

The photovoltaic effect was found in α-Si:H in 1974.[89] Since then, various solar cell designs have been studied, including p-n homojunction, p-i-n (p-type, intrinsic, n-type), Schottky barrier, MIS, and SIS cells. A typical Schottky barrier design consists of depositing α-Si:H by glow discharge in a 0.01–0.05-μm n-type phosphorus-doped layer on a steel substrate followed by an intrinsic undoped layer of 0.3–1 μm. A metal such as platinum with a high work function is then evaporated onto the surface to about 0.005 μm thickness, and finally an antireflection layer of about 0.45 μm of ZrO_2 is evaporated over the metal.[90]

The MIS structure is obtained by depositing a 0.002-μm layer of SiO_2 on the intrinsic α-Si:H before deposition of the metal. The highest solar conversion efficiency reported in the scientific literature for α-Si:H is 5.5% in the MIS cell.[90] Open circuit voltages (V_{max}) of about 0.8 V, short circuit currents (I_{sc}) of 10–12 mA/cm², and fill factors of 0.5–0.6 are typical of state-of-the-art cells under AM1 exposure conditions. A 5.5% efficiency has been obtained also for the *p-i-n* structure for which a 0.02-μm *p*-type layer was deposited on stainless steel followed by a 0.5-μm undoped layer, a 0.008-μm *n*-type layer, and a top layer of indium tin oxide about 0.007 μm thick. Electrical contacts of Ti/Al were evaporated onto the surface.[91] A conversion efficiency of 6.3–10% has been reported without technical detail, other than the fact that a Schottky barrier was used and the composition of the cell included hydrogen and fluorine.[92,93] Indeed, the properties of α-Si may be altered considerably by adding species such as fluorine and oxygen.[94,95] The conductivity of α-Si:F:H deposited in a glow discharge atmosphere of 0.1% PH_3, 90% SiF_4, and about 10% H_2 has reached 10 Ω^{-1} cm^{-1}, as compared to 10^{-12} Ω^{-1} cm^{-1} for intrinsic α-Si:H and 10^{-2} Ω^{-1} cm^{-1} for phosphorus-doped α-Si:H.[96]

Because the study of how to influence and control the properties of amorphous silicon alloys is in its infancy, it is difficult to assess the potential for development of these materials as thin film solar cells. Such factors as substrate temperature, partial pressures of reactive gases, contaminants, glow discharge power density, flow of gases through the chamber, and chamber geometry influence the solid state properties of the material deposited in the glow discharge preparation of α-Si:H. The remarkable rate of progress to 5–6% solar conversion efficiencies in the past 5 yr points to further significant progress. Deposition rates of 0.01–1 μm/

min have been achieved, which would indicate that large-scale, inexpensive deposition processes may be developed to produce these very thin solar cells. Furthermore, clever methods are being developed to prepare large-area arrays of cells connected in series without the necessity for the normal front surface contact grids.[97] These monolithic solar cell panels of α-Si:H are schematically illustrated in figure 4.10. They can be made by photolithographic techniques to provide the appropriate integral circuitry among side-by-side, long and narrow cells. Basically, a transparent conductive oxide, such as indium tin oxide (ITO), on a substrate such as glass is divided into long, narrow, parallel strips followed by deposition of the appropriate conductive and semiconductive layers of α-Si:H in continuous layers over the strips. The α-Si:H layers that may form *p-i-n* junctions are then divided as was the ITO, but with the dividing mask shifted sideways slightly with respect to the divisions in the ITO. The backside electrode is then deposited in such a way as to

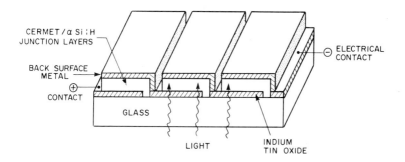

Figure 4.10 Schematic representation of amorphous silicon solar cells arranged in series connection using photolithographic techniques during materials deposition. (Printed with permission of Pergamon Press from reference 97)

make contact with the α-Si:H surface and the exposed ITO in the grooves. With proper technique, it provides a series interconnection between the front electrode of one cell and the back electrode of the adjacent cell without shorting between the back electrodes of adjacent cells. The backside electrode covers one of the walls of the α-Si:H in each separation, but because of the low lateral conductivity of these layers and the thinness of the cells, this is not a serious loss. These types of cells have been made in 36-cm² arrays consisting of nine cells 0.68 cm wide and 5.9 cm long separated by spaces of 0.02 cm. The efficiency was 2.6% with a V_{max} of 6.5 V.[97] This configuration offers the advantage that the inactive area due to front surface grid connection is reduced from about 10% in typical cells to about 1%. There is no basic limit to the device size, and internal power losses should be minimized by operating at low currents and high voltages. The combination of automated, continuous manufacturing processes possible with thin film deposition and the unique solid state properties of amorphous silicon alloys portend major reductions in the cost of making photovoltaic arrays in the future. Considerable scientific attention is expected to continue in this approach to making stable solar cells of more than 10% conversion efficiency.

Cadmium Sulfide-Cuprous Sulfide Heterojunctions

The oldest and most studied thin film heterojunction solar cells are CdS/Cu₂S. These cells are typically lower in solar conversion efficiency than large-crystal silicon-based cells but offer the advantage of lower cost. The photovoltaic effect was first observed in this material in 1954,[98] and development since then has resulted in devices with conversion efficiencies up to 9.1%.[99] Cadmium sulfide is a semiconductor with a band gap of about 2.4 eV, and cuprous sulfide has a band gap of 1.2 eV. As illustrated in figure 3.6, both materials have the potential for yielding solar conversion efficiencies of about 25–27%. When they are grown together to form a heterojunction, an energetic band structure is formed similar to that shown schematically in figure 4.11.[100] The exact nature of the junction depends greatly on the nature of the junction-forming process and doping within the semiconductors. In this particular cell structure, p-type Cu_2S is joined to intrinsic CdS, which in turn is on n-type CdS. Gold grid contacts are connected to the Cu_2S and zinc to the n-CdS. Flow of photoelectrons from the p-type Cu_2S to the n-type CdS changes the representation of energy levels when the cell is illuminated as shown.

Absorption of light in the Cu_2S, which has a band gap of 1.2 eV, causes the photocell to be photosensitive to light from the ultraviolet to 1.0 μm. Absorption may also occur in the junction itself, where an intermediate material composition is present. Indeed, the spectral response of a cell depends greatly on the nature of the junction. The material at the junction is likely to be an intermediate between stoi-

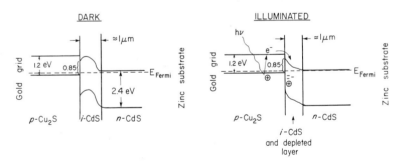

Figure 4.11 Schematic representation of the electronic band structure in CdS/Cu₂S heterojunction solar cells. (Printed with permission of the Institute of Electrical and Electronic Engineers, Inc., from reference 100)

chiometric CdS and Cu_2S, yielding absorption dependent upon the process of forming the junction. Good matches to the AM1 spectrum can be obtained.[101] Because both CdS and Cu_2S are strongly absorbing materials compared with Si (figure 3.3), light may be absorbed in a very short distance. This allows these cells to be much thinner than silicon cells, and material can be conserved. However, the junction needs to be close to the surface and high surface layer photocarrier recombination losses may be encountered. Although large single crystals of CdS can be melt grown, much like silicon, and then treated with p-typing cuprous ions and used as a solar cell, much easier processes are available. By far the most used technique for preparing these cells is vapor deposition of CdS followed by conversion of the surface layer to Cu_2S by dipping it into an aqueous solution of cuprous chloride. Typically, the CdS is vacuum deposited onto a 25–100-μm-thick Kapton polyimide flexible film that has been previously coated with silver plus polyimide varnish and plated with zinc. The CdS layer, which is about 5–50 μm thick for optimal performance,[102] is then bathed in an acidic solution of cuprous chloride or otherwise treated[103] by sputtering or flash evaporation of Cu_2S to an optimal thickness of 3.5 μm or less.[104,105] A metal such as gold is then evaporated onto the Cu_2S layer to provide an electrical contact. It may be thin enough to be effectively transparent or it can be prepared as a grid of contacts. The surface is then covered with a transparent adhesive that adheres a 25–100-μm-thick Kapton polyimide protective cover. A diagram of a typical structure is shown in figure 4.12. The resulting solar cell is polycrystalline, flexible, and protected by front and back layers of polyimide. Present-day technology provides commercially available vacuum deposition units with inlet and outlet ports for continuous flow of material through the chamber. Hence, one might visualize a continuous process

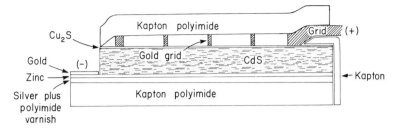

Figure 4.12 Typical construction of a CdS/Cu₂S heterojunction solar cell.

of zinc plating on a plastic base that has contact electrodes attached, CdS deposition, Cu$_2$S formation by dipping, vacuum deposition of a metal grid, and application of a protective covering with metal contacts. This process is schematically described in figure 4.13. This, of course, is a very simplified view of a complicated, but possibly very productive, manufacturing line for producing low-cost cells.

Other processes have been developed for making these types of cells, notably, a dry process involving evaporation of CuCl onto CdS followed by annealing that facilitates an exchange of cations within the sulfide lattice structure, yielding a Cu$_2$S/CdS junction.[106] Extremely thin layers of Cu$_2$S (0.1–0.3 μm) yield relatively high efficiency cells based upon photons absorbed, but light absorption is diminished relative to thicker layers, so that overall cell efficiency remains low. Texturing of the CdS surface or other means of scattering light through the layer can be beneficial.

Because the evaporated CdS is polycrystalline, CdS/Cu$_2$S cells suffer from the same types of surface and grain boundary photocarrier losses as mentioned above for polycrystalline-silicon cells. Figure 4.14 schematically illustrates the cross section of cells. To provide strength and integrity of the CdS layer and to avoid pinholes that cause short circuits, the thickness of the CdS layer is typically 20–

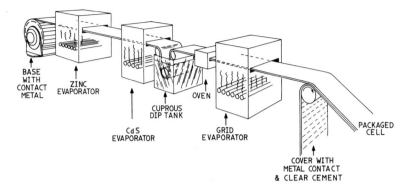

BASE
WITH
CONTACT
METAL

ZINC
EVAPORATOR

CdS
EVAPORATOR

CUPROUS
DIP TANK

OVEN

GRID
EVAPORATOR

COVER WITH
METAL CONTACT
& CLEAR CEMENT

PACKAGED
CELL

Figure 4.13 Schematic representation of a continuous process for manufacturing CdS/Cu$_2$S solar cells.

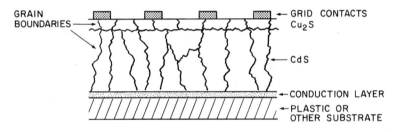

GRAIN
BOUNDARIES

GRID CONTACTS

Cu$_2$S

CdS

CONDUCTION LAYER
PLASTIC OR
OTHER SUBSTRATE

Figure 4.14 Schematic representation of the grain boundaries in a polycrystalline thin film CdS/Cu$_2$S solar cell.

30 μm. This is considered too thick for appropriate conservation of materials, particularly with cadmium, which is relatively scarce.[12] Pyrolytic chemical spraying of the layers onto glass in-line with glass manufacturing has been proposed to yield 2–3-μm CdS layers in a very low cost process.[107] Electrophoretic deposition of CdS from an aqueous sol has been demonstrated to give 1–3-μm thick layers of good quality.[108] Deposition of about 0.01-μm CdS particles into a 1.5-μm layer on stainless steel followed by annealing, evaporation of 0.1-μm Cu_2S, and screen printing gold or silver contacts in a grid has given cells with 1–4.7% efficiency. If the conditions of such a process could be found that produced 10% or higher efficiencies, a very low cost process for manufacturing could be envisioned.[108]

Values of V_{max} in state-of-the-art CdS/Cu_2S cells range from 0.45 to 0.52 V.[109,110] The short circuit currents range from 20 to 24.8 mA/cm², and efficiencies in AM1 light have been reported up to 9.2%.[48] Because of the particular relation of the energy bands between CdS and Cu_2S, efficiencies above 15% are difficult to envision. One of the principal scientific thrusts to improve the efficiency of "cadmium sulfide" solar cells is to substitute other semiconductors such as CdTe, InP, or $CuInSe_2$ for the cuprous sulfide.

Novel Heterojunctions

This category of solar cells refers to perhaps fifty selections of materials in various formats that have been reported to yield from 2 to 22% solar conversion efficiency. They are those that have not been studied and developed to the extent that Si and CdS/Cu_2S cells have been, but have been studied sufficiently to be considered of reasonable potential for future development. One very promising type—that based upon gallium arsenide—will be considered in specific detail later in this chapter because of its special properties that make it particularly useful for converting high-intensity light to electricity.

Several heterojunction solar cells are based upon inorganic semiconductors combined with CdS. Materials with band gaps of 1–1.6 eV, low resistivity, and a lattice structure similar to CdS are favorable. Companion materials include CdTe, $CuInSe_2$, InP, $CuInS_2$, CdZnS, Cu_2Te, ZnTe, and ZnSe. Various formats of semiconductor-semiconductor, Schottky barrier, MIS, and SIS cells have been studied. Another category is based upon GaAs. Gallium arsenide can be p-typed with Ge or Zn, and n-typed with Sn, Te, or Si to yield high-efficiency homojunction cells. It can be combined with $Ga_{1-x}Al_xAs$, GaAlSb, AlAs, or InGaP to form various heterojunction formats. Various metals have been used with GaAs to form Schottky barrier cells. Other photovoltaic cells have been based upon InSnO/CdTe, Cu_2O, Zn_3P_2, $ZnSiAs_2$, $CuGaSe_2$, $AgGaSe_2$, and a myriad of compound semiconductors. The performance of these materials has been reviewed.[3–15] As a generalization, solar cells based on a given junction have higher efficiency when prepared in large-crystal layers than in polycrystalline thin films. In large-crystal layers, heterojunctions of CdS with CdTe, InP, and $CuInSe_2$ have been made with 10–15% conversion efficiencies.[111–113] Homojunction GaAs cells have reached 19%,[114] and a heterostructure p-type GaAlAs over the p-layer of a p-n junction GaAs single crystal has yielded 22% efficiency.[115] Such high-efficiency cells are very promising for special applications for which high efficiency is required. However, the techniques involved in preparing these materials make them very expensive. When less expensive processes are used and polycrystalline thin-film GaAs cells are prepared, 5–6% efficiency is more typical.[116] A relatively promising low-cost process for making n-CdS/n-CdTe/p-CdTe/p^+-Cu$_2$Te cells on a ceramic base has yielded cells with more than 8% efficiency.[117] Another thin film "cadmium sulfide" cell based upon evaporative disposition of all the cell layers in a single vacuum chamber has reached 9.4% solar conversion efficiency.[118] This cell consists of a substrate, a metal back contact, about 2.5 μm CuInSe$_2$, about 0.5 μm CdS, about 1.5 μm indium-doped CdS, and a front surface aluminum grid contact. Thin film heterojunctions between CdSe and Se also show promise.[119] Because of the wide variety of ways to form photovoltaic cells based upon inorganic semiconductors, there is great optimism in the scientific community that several alternative means for making cells in large scale at low cost will be forthcoming.

Organic materials with semiconductive properties offer hope for making low-cost, easily fabricated, thin film photovoltaic cells;[120] however, their electrical properties militate against high conversion efficiencies. Some cells based upon the Schottky barrier have shown a V_{max} of 0.5–1 V. Although these voltages are in the same range as those of inorganic semiconductors, the associated photocurrents are typically very small. Conversion efficiency of about 10^{-4}% has been reported for a tetracene film sandwiched between aluminum and gold electrodes.[121,122] Magnesium phthalocyanine between aluminum and silver yields an efficiency of 10^{-2}%.[123] Chlorophyll, various phthalocyanides, squarylium dyes, and a complex of poly(N-vinylcarbazole) with trinitrofluorenone have been studied.[124–126] Various problems of high resistivity, space charge limitations, and uncertain band structures have limited the efficiencies to less than 0.1%. Before the organic semiconductors compete favorably with Si, α-Si:H, or CdS, improvement of several orders of magnitude will have to be made in respect to lower resistance and higher current flow. Because there is little molecule-to-molecule or atom-to-atom coupling in the molecular organic crystals relative to the ionic inorganic crystals, one would expect that they would have poorer transport of energy, lower photocarrier mobilities, sharper absorption peaks, and more luminescence losses than the inorganic semiconductors. Hence, it appears that organic semiconductors provide little immediate encouragement with regard to yielding high-efficiency solar cells.

Photoelectrochemical, Photocatalytic, and Photogalvanic Solar Cells

In addition to the use of homojunctions and heterojunctions in absorptive semiconductive materials as a means of separating photogenerated holes and electrons, one may also use solutions and semiconductive electrodes. In photocatalytic cells, a chemical reaction such as decomposition of water is caused by light absorption followed by photooxidation and photoreduction. The photoproducts, oxygen and hydrogen in this case, can serve as energy storage and later be utilized as energy sources by burning or by chemical reactions such as in a fuel cell to form electricity. Because the energy difference between the initial (H_2O) and final (O_2 and H_2) states is only 1.23 eV, photons of wavelength shorter than 1.0 μm, about 80% of the solar energy spectrum, can be used to catalyze decomposition.

In photoelectrochemical and photogalvanic cells, no net chemical products are created, but electricity can be provided in an external circuit. A photoelectrochemical cell is typically made by dipping one semiconductor electrode into an aqueous solution and electrically connecting it to a metal electrode inserted into the same solution. Because of energetic band bending within the semiconductor, absorption of light *in the semiconductor electrode* results in photocarrier separation and a current flow through an external circuit. A photogalvanic cell can be produced by inserting two connected metal electrodes into a solution of material that, when illuminated in the vicinity of one electrode, will generate charge carriers that produce electrical power by back reacting through the external circuit. Hence, in a photogalvanic cell, light absorbed *in the electrolyte* gives rise to photocarrier separation and flow of current through an external circuit. A photoelectrochemical cell of the appropriate energy level design and electrolyte selection may be a photocatalytic cell because it produces chemical products at the electrode, instead of causing a current in the external load. Photogalvanic cells of the appropriate design may serve as storage batteries that are recharged by shining light on them.

In discussing these cells and their mechanisms of operation, let us begin with photoelectrochemical cells, then pass to photocatalytic and finally photogalvanic cells. A photoelectrochemical cell is analogous to an MIS solar cell in many ways (see figure 4.6). In the metal-insulator-semiconductor cell, one polarity of the photocarriers separated in the semiconductor passes through a thin insulative layer to recombine with the opposite polarity charge that has completed the circuit through an external load and back to the metal. If we replace the insulator with a liquid solution, a photoelectrochemical cell is formed in which one polarity of the photocarriers separated in the semiconductor passes through the electrolyte and recombines with the opposite polarity charge that has completed the circuit through an external load and back to the metal. The energy band diagram of a hypothetical cell is shown in figure 4.15. The top part of the figure shows the flat band energies of an *n*-type semiconductor, an electrolyte, and a metal. The bottom part shows the equilibrated band structure after the semiconduc-

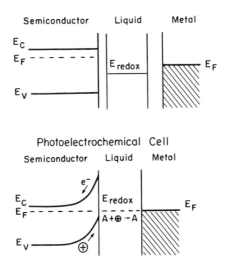

Figure 4.15 Electronic band diagram of a hypothetical photoelectrochemical cell.

tor and the metal (externally connected through a load) have been immersed in the solution. Because of absorbed electrolyte ions at the solution interface with the semiconductor (the Helmholtz layer), a thin barrier to hole transport to the electrolyte may be envisioned, much like the thin oxide barrier in the MIS cell (figure 4.6).[127] The reduction-oxidation potentials of the solution can be chosen to optimize their relation to the Fermi level of the semiconductor, thereby maximizing the potential for photocarrier separation and transport away from the junction. A schematic diagram of a photoelectrochemical cell construction is illustrated in figure 4.16 along with the charge flow during illumination. Obviously, the thin metal electrode pictured could be replaced by a conductive oxide such as indium tin oxide as used in SIS cells.

The efficiency and output voltage characteristics of photoelectrochemical cells depend upon the degree of band

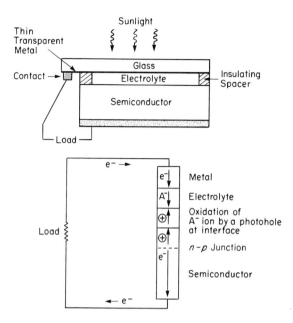

Figure 4.16 Typical construction of a photoelectrochemical cell and a circuit diagram for electrical conduction during illumination.

bending under the surface of the semiconductor and the band gap in relation to the solar energy spectrum. The ideal band gap is between 1.1 and 1.4 eV. Theoretical efficiencies equivalent to those of Schottky barrier cells are expected.[128] For CdTe electrodes with a band gap of 1.4 eV, V_{max} of 0.7 V and 5% efficiency have been obtained.[128] However, in this case the CdTe photoanode decomposes at the liquid surface by the reaction

$$Cd^{++}Te^{--}_{(solid)} - 2e^{-} \xrightarrow{H_2O} Cd^{++}_{(aq)} + Te_{(solid)}.$$

The CdTe surface soon becomes covered with tellurium as the cell operates. This can be alleviated somewhat by adding the redox-active species, Te/Te^{--} in this case, to the solution. A redox electrolyte that is oxidized (accepts holes) more rapidly than the lattice tellurium ions are oxidized is required. Recently it was reported that Ellis, Kaiser, and Wrighton at the Massachusetts Institute of Technology invented an improved cell with stable electrodes, including CdS and CdSe, that had efficiencies of 15%.[129]

Other semiconductive materials in photoelectrochemical cells such as $SrTiO_3$ and TiO_2 are stable[130,131] but have large band gaps of 3.2 and 3.0 eV, respectively.[132] Because they absorb less than 10% of the solar spectrum, they do not produce high conversion efficiencies. Various attempts have been made to form cells from lower-band gap semiconductors, multilayer semiconductors of graduated band gaps, spectral sensitization, and thin metal protection for the semiconductor, but most have been plagued by decomposition of the surface. However, intense scientific effort is going toward developing materials that will be stable and easy to make.[132–134] Various photoelectrochemical cells have been made recently with solar conversion efficiencies of 8–9% with stabilities corresponding to three or more years as a solar transducer.[135,136] Electrodes such as zinc have been incorporated to combine photoelectrochemical generation with storage of electricity.[137] Such a combination may yield manufacturing efficiencies and savings of space required for electrical generation and storage.

A photoelectrochemical cell can be specially designed to be a photocatalytic cell. When the semiconductor is selected to have the appropriate work function relative to its band gap and the redox potentials for decomposition of the solution into stable products, a decomposition can be photocatalyzed by the absorption of light in the semiconductor. For example, if an aqueous electrolyte is used, the decomposition of water into H_2 and O_2 can be stimulated by light. The photohole at the semiconductor interface with the water must have the right energy to cause the following reaction in basic water solution:

$$4\oplus + 4\,OH^- \longrightarrow 2H_2O + O_2\!\uparrow.$$

If the work function of the illuminated semiconductor is lower than the potential relative to vacuum for the reaction

$$2e^- + 2H_2O \longrightarrow H_2\!\uparrow + 2OH^-,$$

then the potential at the cathodic metal (connected through the external load) will be sufficient to reduce hydrogen ions to H_2. These conditions are illustrated in figure 4.17 for TiO_2 and $SrTiO_3$ photocatalytic cells.[134] All energy levels are referenced to the standard calomel electrode (SCE), and a pH 14 water solution is assumed with platinum cathodes. Under illumination of the $SrTiO_3$ anode, the electron Fermi energy E_{Fn} becomes more negative than that required to form H_2 from H_2O; thus, hydrogen is bubbled from the platinum-water interface. The pseudo-Fermi energy for photoholes, E_{Fp}, is more positive in figure 4.17 than the potential for oxidation of water to O_2; thus, oxygen is bubbled from the $SrTiO_3$-water interface. It is prudent to put a salt bridge or some other ion permeable barrier between the two electrodes so that the O_2 does not diffuse to the Pt electrode, where it can be reduced again. One may consider the O_2/OH^- redox couple to be in "contact" with the valence band, and the H_2/H_2O couple to be a contact to the conduction band. Hence, the maximum photovoltage achievable from this cell is the difference between these redox couples, 1.23 V. Then the power output that goes to produce O_2 and H_2 is the photocurrent multiplied by 1.23 V.

In the TiO_2-water-Pt cell, the relation of E_{Fn} to the poten-

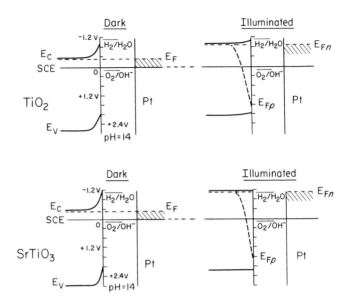

Figure 4.17 Schematic representation of the electronic band structures in photocatalytic cells using TiO_2 and $SrTiO_3$ as the light-absorbing semiconductors, alkaline water electrolyte, and platinum metal cathodes. (Printed with permission of Pergamon Press from reference 134)

tial for the H_2/H_2O redox couple under illumination dictates that H_2 will not be formed (figure 4.17). Hence the cell will not operate to produce H_2 and O_2. This condition can be alleviated by providing a negative bias of about 0.15 V on the platinum electrode.[134]

The production of stable products of the photocatalytic cell that can be stored and later used to produce heat or electricity is very appealing in principle because of the non-coincidence of peak energy demand and peak solar incidence. Of course, electricity can be stored by connecting solar cells to charge batteries or spin flywheels that later produce electricity on demand. There is an economic appeal, however, to directly forming fuels that are energy rich and can be stored easily.

Because of the particular energetics involved in electrolysis of water, the selection of semiconductors to cause decomposition is limited. One novel approach to providing electricity from photocatalytic cells is shown schematically in figure 4.18.[138,139] This is a doped-silicon semiconductor—aqueous HBr—metal photocatalytic cell that takes advantage of the efficient p-n junction in silicon and the reversible reaction

$$2HBr + \text{electricity} \underset{}{\overset{H_2O}{\rightleftharpoons}} H_2\uparrow + Br_2(aq).$$

At small silicon spheres with n-type centers and p-type shells, the following anodic reaction occurs upon illumination:

$$2\oplus + 2HBr \xrightarrow{H_2O} Br_2(aq) + 2H^+.$$

At small silicon spheres with p-type centers and n-type shells, the following cathodic reaction is catalyzed by light absorption:

$$2e^- + 2H^+ \xrightarrow{H_2O} H_2\uparrow.$$

The gaseous hydrogen reaction product is stored in the interstices of a metal alloy as a hydride, and the dissolved liquid bromine is stored in a closed vessel, which may be used also as a heat exchanger to cool the photocatalytic cell and provide heat for comfort purposes. The H_2 and Br_2, representing chemically stored energy, can be recombined in a fuel cell to produce electricity and hydrobromic acid, which is recirculated to the solar collector as shown in the figure. A separator is provided within the collector to prevent direct recombination of H_2 and Br_2 in a photogalvanic re-

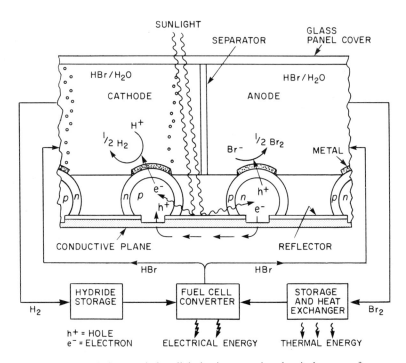

Figure 4.18 Novel photocatalytic cell design incorporating chemical storage of energy and a fuel cell for electrical power generation. [Reprinted from *Electronics* (6 November 1980). Copyright © 1980. McGraw-Hill, Inc. All rights reserved]

action. Electrical connection is made between the silicon
spheres with p- and n-type centers, respectively, by a back
conductive plate. The circuit can be completed through the
liquid via the external flows through the fuel cell or by ionic
exchange through the liquid separator in the solar collector.
The energy band profiles are schematically presented in figure 4.19. By connecting the n-type cores to the p-type centers through the conductive backing, if the V_{max} across the
junction between centers and shells approaches 0.6 V, then
the voltage difference across the liquid can approach 1.2 V.
The difference in potentials between the H^+/H_2 and Br^-/Br_2
redox couples in aqueous solution is about 1 V.

This type of photocatalytic cell has achieved a 13% overall
efficiency,[139] even in arrays,[138] and cost effective means are
used for producing the silicon p-n junction cell components.
Spheres can be made with no silicon waste by forcing molten
silicon through an orifice and solidifying. These are cast into
a glass matrix with a conductive backing layer. No additional
electrical contacts such as grids are required. Thus, with

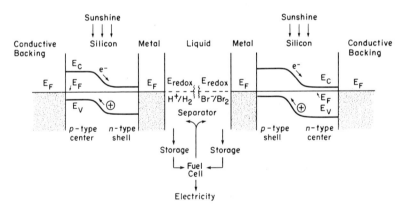

Figure 4.19 Schematic representation of the electronic band structure in the photocatalytic cell illustrated in figure 4.18.

appropriate sizing of the hydrogen and bromine storage, this system of converting solar energy to electricity may provide electricity on demand without the intermittency effects due to variations in sunshine.

Besides the photoelectrochemical and photocatalytic cells, photogalvanic cells can also be constructed by dipping two appropriate electrodes into a photoreactive solution. In a photogalvanic cell, incident light is absorbed in an absorbing liquid to generate photocharge carriers that can either slowly recombine directly (a loss process) or diffuse to opposite electrodes, where they effectively back react through the external circuit.[140] Control of the reaction dynamics in solution is important and can sometimes be accomplished by phase separation.[141] However, because of the many solution loss processes possible, these types of cells are typically very inefficient. The highest value developed with an iron-thionine cell is less than 0.1%.[142] An attempt to combine such a cell in series with a photoelectrochemical cell with magnesium mesotetraphenylporphyrin as a photoactive electrode failed to increase the efficiency.[143] Because photogalvanic cells appear to offer little potential for development into efficient solar converters, no further discussion on this matter is offered.

Concentration Solar Cells

Because of the cost involved in making large flat plate arrays of solar cells that absorb the rather diffuse influx of sunshine, various schemes have been developed to collect the sunlight and concentrate it onto a small solar cell for conversion to electricity. A simple concept is to take a lens, such as a magnifying glass, and concentrate the sunlight collected over, say, 10 in.^{2} onto a 1-in.^{2} silicon solar cell. If we neglect optical losses, this results in collection of ten times more light onto the solar cell by adding the relatively inexpensive lens. Various concentrating devices have been used, including reflective troughs of various types with solar cells at the bottom, linear and circular Fresnel lenses,[144] parabolic cylinders, circular cylinders, paraboloids, and heliostats, in attempts to drive down the cost of solar energy collection. Of course, the same area must be used to intercept the sunshine, but the material that intercepts and reflects it or focuses it onto the photovoltaic converter may be less expensive on an area basis than the converter.

Most concentrators are effective only with specular light, that is, the light shining directly from the sun onto them and not scattered from clouds, dust, or the environment. Subject to elevation, latitude, solar altitude, solar declination, turbidity of the air, and cloudiness, the proportion of sunlight that reaches the earth's surface is roughly 60% specular in the summer and 40% specular in the winter in the northern United States and southern Canada.[145,146] These percentages are higher in the "sunbelt" areas of the south and southwestern United States; however, a significant fraction of solar energy across the United States is diffuse sky radiation and is not collected effectively in concentration cells. The diffuse-to-specular radiation ratio affects the optimal design of solar tracking systems as well as the optimal tilt on stationary flat solar cell arrays. Solar tracking on a relatively clear day in Iowa may increase the daily power output by about 30% relative to a stationary array.[147] The increase is caused by more collection of the specular energy in the early morning and late afternoon hours. Various models have been used to predict the optimal tilt on a flat, nonconcentrating array.[148,149] In the northern United States, tilting an array 45° from horizontal to the south increases the average yearly collection by 20–30%. Again, the optimal tilt depends not only upon the angle toward the specular light but also the angle toward the upper sky diffuse radiation and reflections in the immediate environment.

When solar concentration sysytems are used, the solar cell material must be selected to withstand higher than ambient temperatures unless active cell cooling is provided. Solar cell efficiency is degraded at higher temperatures particularly, because of lower effective operating voltages. This factor is less important for larger band gap semiconductors such as GaAs (1.4 eV) relative to Si (1.1 eV). Because of the high photocurrents involved, special care is taken to reduce the series resistance of the cell by carefully controlling the junction depth and front surface electrical connections. For a variety of physical and electrical reasons, most presently known heterojunctions of the Schottky barrier, MIS, SIS, and semiconductor-semiconductor types are not candidates for solar cells at the temperatures and photocurrents involved at concentration ratios in the region 50–1000 or higher. Homojunction materials, such as n-p junctions in Si and GaAs, are emphasized, with GaAs being the better material at the highest concentration ratios.[150] Typical costs of silicon devices for use in concentration cells are \$200/ft^2, and those for GaAs cells are \$2000/ft^2.[151] Of course, if concentration ratios of 100 and 1000 were achieved in the respective cases, the photovoltaic cell cost would be reduced to about \$2/ft^2 of collector area. Because this no longer represents the major cost per square foot associated with the collector array, photovoltaic cell cost reduction is less important than ensuring that the cell has a high conversion efficiency.

Because GaAs has the inherent solid state properties to maintain high solar conversion efficiency at high temperatures, considerable research is being devoted to its development for solar concentration cells. Cells have been made that yield 17.2 and 19.1% efficiency under light concentrations of 896 and 1735 suns, respectively.[152] A very high efficiency cell with an epitaxially grown layer of $Ga_{1-x}Al_xAs$ over GaAs has been produced to yield a shallow, low series-resistance, homojunction cell, as illustrated in figure 4.20. This type of cell has operated at 23% efficiency at a concentration of 10 suns and 19.1% at 1700 suns, during which it was heated to 60°C.[153] Efficiencies of 28–29% for a 1000 AM1 sun concentration at 25°C and 15% at 350°C are theoretically possible with such cells.

Silicon solar cells can be utilized in concentration cells most effectively if they are specially designed to have low resistivity and increased optical absorption of light relative to the normal *p-n* junction silicon cells. Single-junction cells are useful up to perhaps a concentration of 100 suns, but above that, multijunction cells must be designed. Typically, very fine grids and multiple terminal contacts are made to reduce the series resistance. Single-junction devices have been made that reach 15.5% efficiency at a concentration of 23 suns and 12.8% at a concentration of 109 suns.[154] Typi-

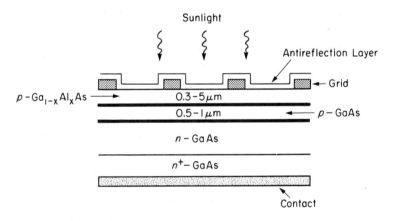

Figure 4.20 Structure of a high-efficiency gallium arsenide homojunction solar cell for use in solar concentration arrays.

cally, the efficiency degrades by a percentage point for every 16–22°C increase in temperature.[155]

Various types of multijunction silicon cells have been designed for use at solar concentrations of over 100. Vertical junctions formed by 250-μm p^+-n-n^+ slices bonded at 700°C with aluminum foil between slices have yielded efficiencies of 6.1% at a concentration of 329 suns and 130°C.[156,157] Horizontal multijunction structures have also been prepared with alternating p- and n-type layers of about 2 μm between the surface p^+ and back n^+ layers. Because of the multiple internal junctions, photocarriers are efficiently separated and, as long as the layers are thin, sufficient conduction can occur across the cell to yield reasonable efficiencies. A six-layer structure has been reported that has an 11.1% conversion efficiency at a concentration of 623 suns at 56°C.[158] Another type of multijunction silicon cell is shown in figure 4.21. In this cell, alternating p^+- and n^+-type stripes are prepared in the "back" surface of a high-quality silicon crystal. This should be a crystal that permits a long lifetime for photocarriers. The p^+ and n^+ stripes are electrically connected so as to form the cell contacts. Light enters the "front" of the cell and provides photocarriers that are separated by the fields at the junctions on the back surface. This type of cell has given a 16.5% conversion efficiency under a concentration of 220 suns at 15°C.[159]

Another type of concentration cell is called the thermophotovoltaic cell.[160–162] In this cell a large concentration ratio up to 10,000 is used. The light is directed into a blackbody absorber-radiator sphere through a hole in the sphere, as illustrated in figure 4.22.[162] The solar energy is absorbed in a black incandescent absorber-radiator integrating sphere designed to operate at about 2000K. At this temperature the peak intensity of radiation is at about 1 μm, for which the energy conversion to electricity in silicon cells is most effi-

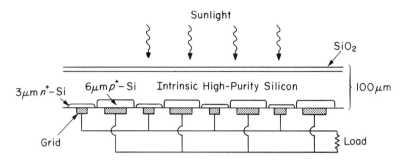

Figure 4.21 Multijunction silicon solar cell designed for use in solar concentration arrays.

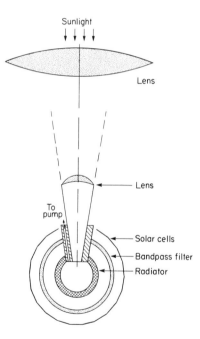

Figure 4.22 Typical structure of a thermophotovoltaic cell. (Printed with permission of Elsevier Sequoia S.A. from reference 162)

117 State of the Art in Photovoltaics

cient. The radiation between 0.6 and 1.1 μm is allowed through a filter and reaches an outer sphere of silicon solar cells. That radiation that falls outside of the 0.6–1.1-μm band is reflected back to the absorber-radiator, where it reheats the sphere. Special precautions against conductive and radiative heat losses are taken in the selection of materials and the design of the sphere. In the converter illustrated, a vacuum is maintained in the two inner spheres. Such converters are expected to have conversion efficiencies in the 20–50% region, but few have actually been built.

Yet another novel concentration cell has been proposed to achieve very high efficiencies.[163] In this cell the impinging solar light is separated into separate bands of energy by a chromatic dispersion concentrator. These bands of light are then directed to different single-junction solar cells that are matched in band gap or absorption to the particular band of the solar light directed to them. This minimizes the losses in a given type of cell due to heating by photons with higher energy than the band gap. Using four bands of light matched to four solar cell materials with band gaps of 0.7, 1.1, 1.45, and 2.4 eV yields a theoretical efficiency of 26%. Depending upon the selection of cell material, efficiencies of 25–35% are expected, with 28% having been demonstrated recently.[164]

Under the auspices of the US Department of Energy, various industrial companies are trying to reduce the cost of photovoltaic energy conversion by making practical concentration cells. For single-junction cells designed to operate in a concentration of 50–1000 suns, silicon cells have yielded 18–20% efficiency, and AlGaAs cells have reached 23% efficiency.[165] Optical concentrators have been demonstrated with about an 85% efficiency, and overall array efficiencies of 8–12% have been achieved. Currently the price of an array is about $100/ft^2. Projected array efficiencies are 15–20% with prices of $10–15/ft^2. Using the following equation for conversion to price per peak watt, $/W_p = $/ft^2 \div (100$ $W/ft^2 \times$ fraction converted), the projected range of costs is $0.5–$1.0/W_p. This is about one tenth the present price of solar cell arrays of both the flat plate and the concentration types.

Summary of Key Factors in Solar Cells

The three most important factors inherent in solar cells that govern their widespread application to electricity production are cost, efficiency, and stability. The state of the art in researches on efficiency and a discussion of possible future gains have been presented in the preceding sections for different types of cells. Also, reference has been made to current costs associated with various cells and the potential for cost reduction through new cell design and improved manufacturing processes. However, relatively little discussion has been devoted to cell stability. It is generally regarded that a solar cell must perform at least 20 yr at an efficiency that is at least 10% and does not decline significantly. Of course, the normal reduction in efficiency due to blockage of light (by, among other things, dirt, bird droppings, snow, leaves, and fungus growth) on the surface of the array should be reversible by cleaning. Reduction in photovoltaic conversion efficiency (by, for example, thermal stress, reactions with moisture or environmental chemicals, and increased resistance in electrical contacts) that may not be recovered by minimal maintenance is detrimental to the reliability and economical use of solar cells.

Of all the cells described in this chapter, only large-crystal silicon cells have been shown repeatedly to have the required stability in a terrestrial setting. Of course, silicon is the material that has undergone the greatest development and is most readily available in commercial arrays. The homojunction large-crystal *p-n* homojunction silicon cell has a high melting point, and diffusion in it of dopant materials is insignificant at ambient conditions under unfocused direct sunlight. Hence, if sturdy and reliable electrical contacts are made to the silicon, these cells are stable for more than 20 yr. Polycrystalline homojunction silicon and homojunction GaAs cells are expected to have adequate stability, but have not been developed fully enough to establish with certainty that a 20-yr lifetime is feasible under operating conditions. Cells based upon α-Si:H have not been made with adequate stability. This may be caused by hydrogen evolution at high temperatures,[166] but more likely it is caused by interaction with water vapor that can adversely affect the performance of Schottky barrier, MIS, and SIS cells.[167,168] Most heterojunction cells of the Schottky barrier, MIS, and SIS types have to be very carefully designed and protected from the environment to provide an extended lifetime.[169] This may be caused by problems such as impurity reaction at the junction to change the solid state conduction characteristics of the cell, oxidation of metal contacts, growth of the insulative oxide layer in MIS cells, reaction of the semiconductor with the metal at high temperatures, diffusion of one material into another, and loss of intimate junction contact because of differing materials expansion upon heating and cooling.

The CdS/Cu$_2$S heterojunction solar cell typically degrades in performance with time and has been studied extensively over the last 15 yr to overcome this problem. The CdS/Cu$_2$S cell instability is ascribed to the polycrystalline structure, which presents a high effective surface area for reactions with chemical contaminants; the fact that cuprous sulfide has a phase transformation near the normal operating temperature and considerable copper ionic mobility; and stoichiometric changes from oxidation of the Cu$_2$S layer. The combination of heat and illumination decreases the output efficiency.[170] Although prolonged illumination of cells at 25°C to light intensities equivalent to slightly more than direct sunlight causes some degradation of V_{max}, it is reversible by holding the cells in the dark or short-circuiting them.[171] Irreversible degradation occurs when the cells are illuminated and heated above 60°C. This is associated with an increase in ionic conductivity and $\gamma \rightarrow \beta$ phase transition in the region of 80°C. The degradation of CdS/Cu$_2$S cells is much more rapid at high humidity and is a strong function of the electrical load. Under open circuit conditions and illumination, the power output may drop to about two thirds of its original power in a few days.[171] However, under lighter load conditions, this does not happen as readily. The cause of the power loss is thought to be the degradation of Cu$_2$S to CuS and Cu.[172] This reaction has an electrode potential at 25°C of -0.39 V. When the applied voltage across the surface layer of a CdS/Cu$_2$S cell exceeds this, copper is produced that shorts the cell. Thus, at open circuit voltages of 0.47 V, the cell rapidly degrades. Fortunately, the condition for maximum power is a voltage of about 0.33 V, far enough below the threshold that degradation will not occur from this reaction at ambient temperatures and proper selection of load resistances. Other slower failures occur, and various means have been developed to help stabilize the cell.[173,174] However, no cells of adequate stability and efficiency have been developed for commercial use. The greatest emphasis over the past few years has been on replacing the Cu$_2$S with CdTe, InP, CuInSe$_2$, or CuInS$_2$, which may result in a new and stable "cadmium sulfide cell," but requiring more complicated manufacturing processes than the simple dipping of a film of CdS into aqueous CuCl to yield the CdS/Cu$_2$S heterojunction by ion exchange.

Of all the approaches to making solar cells, none has satisfied the three main criteria for widespread utilization: efficiency of 10% or greater; stability of 20 yr; and a cost of below $1/W_p$. The last factor can be converted to cost per square foot by the following relation between cost per peak watt, efficiency, and insolation:

$$\text{cost/ft}^2 \leq \text{cost/W}_p \cdot \text{insolation} \cdot \text{efficiency}$$
$$\leq \$1 \cdot 100 \text{ W/ft}^2 \cdot 0.10$$
$$\leq \$10.00.$$

The cost per square foot may be doubled if the efficiency is doubled in the process. The large-crystal silicon cell arrays and concentration cells come closest to meeting the criteria. Commercially available arrays easily exceed 10% efficiency and meet the stability criterion. However, array costs are currently about $10/W_p$. Commercially available cadmium sulfide solar arrays have only about a 4% efficiency, below the required stability, and cost about the same as silicon cells. The other cells mentioned in this chapter have not been developed to the stage of commercial availability.

References

1. D. M. Chapin, C. S. Fuller, and G. L. Pearson, *J. Appl. Phys.* 25 (1954): 676.

2. H. Kelly, "Photovoltaic Power Systems: A Tour through the Alternatives," *Energy II: Use, Conservation, and Supply*, P. H. Abelson and A. L. Hammond, eds. (Washington, DC: Am. Assoc. Adv. Sci., 1978), pp. 151–160.

3. D. Costello and P. Rappaport, "The Technological and Economic Development of Photovoltaics," *Ann. Rev. Energy*, J. M. Hollander, M. K. Simmons, and D. O. Wood, eds. (Palo Alto, CA: Annual Reviews Inc., 1980), pp. 335–356.

4. H. J. Hovel, *Solar Energy* 19 (1977): 605.

5. H. Ehrenreich and J. H. Martin, *Phys. Today* 32 (September 1979): 25.

6. J. Javetski, *Electronics* 52 (July 1979): 105.

7. H. Kelly, *Science* 199 (1978): 634.

8. E. Bucher, *Appl. Phys.* 17 (1978): 1.

9. B. T. Debney and J. R. Knight, *Contemp. Phys.* 19 (1978): 25.

10. E. A. Perez-Albuerne and Yuan-Sheng Tyan, *Science* 208 (1980): 902.

11. C. Currin, K. Ling, E. Ralph, W. Smith, and R. Stirn, "Feasibility of Low Cost Silicon Solar Cells," *Proc. 9th IEEE Photo. Spec. Conf.* (Maryland: IEEE, May 1972).

12. R. Singh and J. D. Leslie, *Solar Energy* 24 (1980): 589.

13. B. Horovitz, *Industry Week* (26 May 1980): 70.

14. J. Javetski, *Electronics* 52(15) (1979): 105.

15. J. B. Brinton, *Electronics* (11 September 1980): 39.

16. J. Lindmayer, "Characteristics of Semi-Crystalline Silicon Solar Cells," *Proc. 13th IEEE Photo. Spec. Conf.* (New York: IEEE, 1978), pp. 1096–1100.

17. J. G. Posa, *Electronics* (11 October 1979): 43.

18. *Solar Age* (September 1980): 20.

19. A. Luque and E. Lorenzo, *Solar Energy* 22 (1979): 187.

20. A. Luque, E. Lorenzo, and J. M. Ruiz, *Solar Energy* 25 (1980): 171.

21. A. Flat and A. G. Milnes, *Solar Energy* 23 (1979): 289.

22. A. Flat and A. G. Milnes, *Solar Energy* 25 (1980): 283.

23. J. Lyman, *Electronics* (11 September 1980): 40.

24. W. R. Cherry, *Proceedings of the 13th Annual Power Sources Conference* (Atlantic City, NJ: IEEE, May 1959).

25. Ryco Laboratories, Final Report No. AFCRL-66-134, 1965.

26. R. Riel, "Large Area Solar Cells Prepared on Silicon Sheet," *Proceedings of the 17th Annual Power Sources Conference* (Atlantic City, NJ: IEEE, May 1963).

27. J. A. Zoutendyk, *Solar Energy* 30 (1980): 249.

28. R. Crabb, "Status Report on Thin Silicon Solar Cells for Flexible Arrays" *Solar Cells* (New York: Gordon and Breach, 1971), pp. 35–50.

29. H. Ehrenreich, *Principal Conclusion of the American Physical Society's Study Group on Solar Photovoltaic Energy Conversion* (New York: Am. Phys. Soc., 1979).

30. J. C. Schwartz, T. Surek, and B. Chalmers, *J. Electronic Mat.* 4 (1975): 225.

31. T. F. Ciszek, *Mat. Res. Bull.* 7 (1972): 731.

32. A. V. Stepanov, *Bull. Acad. Sci. USSR, Phys. Series* 33 (1969): 1826.

33. *Westinghouse Silicon Dendritic Web: Key to Low Cost Automated Production of Solar Cells* (Pittsburgh: Westinghouse R&D Center, CD78-12300, 1980).

34. W. D. Metz and A. L. Hammond, *Solar Energy in America* (Washington, DC: Am. Assoc. Adv. Sci., 1978).

35. L. Lowe, *Electronics* (6 November 1980): 40.

36. J. D. Heaps, R. B. Maciolek, W. B. Harrison, and H. A. Wolner, *Dip-Coating Process*, ERDA/JPL 954356, *Q. Report No. 1* (1975).

37. R. P. Ruth, H. M. Manasevit, J. L. Kenty, L. A. Moudy, W. I. Simpson, and J. J. Yang, *Chemical Vapor Deposition Growth*, ERDA/JPL 954372-76/1, *Q. Report No. 1* (1976).

38. C. D. Grahm, S. Kulkarni, G. T. Noel, D. P. Pope, B. Pratt, and M. Wolf, *Hot Forming of Silicon*, ERDA/JPL 954506-76/1, *Q. Report No. 1* (1976).

39. F. C. Eversteyn, *Philips Res. Rept.* 29 (1976): 45.

40. H. Garfinkel and R. N. Hall, *Floating Substrate Process*, ERDA/JPL 954350-76/1, *First Q. Prog. Rept.* (1976).

41. A. K. Ghosh, T. Feng, and H. P. Maruska, *Solar Cells* 1 (1980): 421.

42. A. K. Ghosh, C. Fishman, and T. Feng, *J. Appl. Phys.* 51 (1980): 446.

43. C. W. White, J. Narayan, and R. T. Young, *Science* 204 (1979): 4161.

44. D. H. Auston, C. M. Surko, T. N. C. Venkatesan, R. E. Slusher, and J. A. Golovchenko, *Appl. Phys. Lett.* 33 (1978): 130.

45. J. C. Wang, R. F. Wood, and P. P. Pronko, *Appl. Phys. Lett.* 33 (1978): 455.

46. P. Baeri, S. U. Campisano, and E. Rimini, *Appl. Phys. Lett.* 33 (1978): 137.

47. R. F. Wood, R. T. Young, R. D. Westbrook, J. Narayan, J. W. Cleland, and W. H. Christie, *Solar Cells* 1 (1980): 381.

48. D. L. Feucht, "Photovoltaic R and D Program Overview," *Proceedings of the DOE Annual Photovoltaics Program Review for Technology and Market Development* (Massachusetts, 1980), Conf-8004101, NTIS, US Dept. of Commerce, Springfield, Virginia.

49. D. R. Lillington and W. G. Townsend, *Appl. Phys. Lett.* 28 (1976): 97.

50. H. C. Card and E. S. Yang, *Appl. Phys. Lett.* 29 (1976): 51.

51. D. L. Pulfrey, *Photovoltaic Power Generation* (New York: Van Nostrand Reinhold, 1978).

52. T. C. McGill and C. A. Mead, *J. Vac. Sci. Technol.* 11 (1974): 122.

53. D. L. Pulfrey ano R. F. McOuat, *Appl. Phys. Lett.* 24 (1974): 167.

54. P. Viktorovitch, G. Kamarinos, and P. Even, *Prog. 12th IEEE Photo. Spec. Conf.* (Baton Rouge, LA: IEEE, 1976), p. 870.

55. G. Cheek and R. Mertens, *Solar Cells* 1 (1980): 405.

56. G. Cheek, N. Inove, S. Goodnick, A. Genis, C. Wilmsen, and J. DuBow, *Appl. Phys. Lett.* 33 (1978): 643.

57. W. G. Thompson and R. L. Anderson, *Solid State Electron.* (1978): 603.

58. T. Feng, A. K. Ghosh, and C. Fishman, *J. Appl. Phys.* 50 (1979): 4972.

59. T. Feng, A. K. Ghosh, and C. Fishman, *Appl. Phys. Lett.* 35 (1979): 266.

60. H. K. Charles, Jr., and A. P. Ariotedjo, *Solar Energy* 24 (1980): 329.

61. K. Rajkanan, W. A. Anderson, and G. Rajeswaren, *IEEE Electron Device Mtg.* (Washington, DC: IEEE, December 1979).

62. J. Shewchan, *Annual Progress Report* (Solar Energy Research Institute, 1980), Contract XS–9–8233–1.

63. R. B. Godfrey and M. A. Green, *Appl. Phys. Lett.* 34 (1979): 790.

64. D. E. Burk, J. B. Dubow, and J. R. Sites, *Device Research Conf.* (Salt Lake City, UT: 1976).

65. M. Marshall, *Electronics* (3 January 1980): 102.

66. L. Olson, *Polysilicon Review Mtg.* (Golden, CO: Solar Energy Research Institute, June 1979).

67. R. B. Godfrey and M. A. Green, *Appl. Phys. Lett.* 33 (1978): 637.

68. J. G. Fossum and F. A. Lindholm, *IEEE Trans. Elec. Dev.* ED-24 (1977): 325.

69. P. Shah, *Solid State Electron.* 18 (1975): 1099.

70. R. J. Soukup, *J. Appl. Phys.* 48 (1977): 440.

71. R. J. Schwartz and M. D. Lammert, *Proc. IEEE Int. Electron Dev. Mtg.* (Washington, DC: IEEE, 1975), p. 353.

72. T. Matsushita and T. Maimine, *Proc. IEEE Int. Electron Dev. Mtg.* (Washington, DC: IEEE, 1975), p. 353.

73. H. Macomber, *Proceedings of the ERDA Semiannual Solar Photovoltaic Program Review Mtg.* (Washington, DC: Energy Research and Development Administration, 1977), p. 68.

74. J. J. Loferski, *Proc. 12th IEEE Photo. Spec Conf.* (New York: IEEE, 1976), p. 957.

75. A. Goetzberger and W. Breubel, *Appl. Phys.* 14 (1977): 123.

76. M. P. Vecchi, *Solar Energy* 22 (1979): 383.

77. M. Wolf, *Energy Conv.* 11 (1971): 63.

78. R. C. Chittick, J. H. Alexander, and H. F. Sterling, *J. Electrochem. Soc.* 116 (1969): 77.

79. G. A. N. Connell and J. R. Pawlik, *Phys. Rev. B* 13 (1976): 787.

80. D. E. Carlson, "Amorphous Silicon," to be published in *Progress in Crystal Growth and Characterization*, 1981.

81. A. J. Lewis, G. A. N. Connell, W. Paul, J. R. Pawlik, and R. J. Temkin, *Proc. Int. Conf. on Tetrahedrally Bonded Amorphous Semiconductors* (New York: American Institute of Physics, 1974), p. 27.

82. W. E. Spear and P. G. LeComber, *Solid State Commun.* 17 (1975): 1193.

83. D. E. Carlson, C. R. Wronski, J. I. Pankove, P. J. Zanzucchi, and D. L. Staebler, *RCA Review* 39 (1977): 211.

84. W. E. Spear, P. G. LeComber, S. Kalbitzer, and G. Muller, *Philos. Mag. B* 39 (1979): 159.

85. M. H. Brodsky, M. A. Frisch, J. F. Ziegler, and W. A. Landford, *Appl. Phys. Lett.* 30 (1977): 561.

86. P. J. Zanzucchi, C. R. Wronski, and D. E. Carlson, *J. Appl. Phys.* 48 (1977): 5227.

87. J. F. Graczyk, *Phys. Stat. Sol.* 55 (1979): 231.

88. A. Barna, P. B. Barna, G. Rodnoczi, L. Toth, and P. Thomas, *Phys. Stat. Sol.* 41 (1977): 81.

89. D. E. Carlson, US Patent No. 4,064,521 (1977).

90. D. E. Carlson, *IEEE Trans. Electron Dev.* ED-24 (1977): 449.

91. D. E. Carlson, R. W. Smith, G. A. Swartz, and A. R. Triano, "5.5% *p-i-n* Amorphous Silicon Solar Cells," *Extended Abstracts*, vol. 80-2 (Hollywood, FL: Electrochemical Society Mtg., October 1980), p. 1428.

92. L. Waller, *Electronics* (28 August 1980): 41.

93. L. Waller, *Electronics* (31 January 1980): 40.

94. S. R. Ovshinsky and A. Madan, *Nature* 276 (1978): 482.

95. M. A. Paesler, D. A. Anderson, E. C. Freeman, G. Moddel, and W. Paul, *Phys. Rev. Lett.* 41 (1978): 1492.

96. A. Madan, S. R. Ovshinsky, and E. Benn, *Philos. Mag.* 40 (1979): 259.

97. J. J. Hanak, *Solar Energy* 23 (1979): 145.

98. D. C. Reynolds, G. Leies, L. L. Antes, and R. E. Masbruger, *Phys. Rev.* 96 (1954): 533.

99. A. Barnett, J. A. Brogagnolo, R. B. Hall, J. E. Phillips, J. D. Meakin, *Proc. 13th IEEE Photo. Spec. Conf.* (New York: IEEE, 1978), p. 419.

100. L. Shiozawa, G. Sullivan, and F. Augustine, *Proc. 7th IEEE Photo. Spec. Conf.* (Pasadena, CA: IEEE, 1968), p. 22.

101. A. N. Casperd and R. Hill, *Solar Cells* 1 (1980): 347.

102. I. Abrahamsohn, US Patent 3,376,163 (2 April 1968).

103. J. David, S. Martinuzzi, F. Cabane-Brouty, J. Sorbier, J. Mathieu, J. Roman, and J. Bretzner, "Structure of CdS-Cu$_2$S Heterojunction Layers," *Solar Cells* (New York: Gordon and Breach, 1971), pp. 81–94.

104. S. Yu. Pavelets and G. A. Fedorus, *Geliotekhnika* 7 (1971): 3.

105. S. Yu. Pavelets and G. A. Fedorus, *Applied Solar Energy* 7 (1973): 1.

106. T. S. teVelde and J. Dieleman, *Philips Res. Rep.* 28 (1973): 573.

107. J. F. Jordan, *Proc. 11th IEEE Photo. Spec. Conf.* (Scottsdale: AZ: IEEE, 1975), p. 508.

108. E. W. Williams, K. Jones, A. J. Griffiths, D. J. Roughley, J. M. Bell, J. H. Steven, M. J. Huson, M. Rhodes, and T. Costich, *Solar Cells* 1 (1980): 357.

109. K. W. Böer, *Proc. 11th IEEE Photo. Spec. Conf.* (Scottsdale, AZ: IEEE, 1975), p. 514.

110. J. D. Meakin, B. Baron, K. W. Böer, L. Burton, W. Devaney, H. Hodley, J. Phillips, A. Rothwarf, G. Storti, and W. Tseng, *6th Int. Solar Energy Soc. Conf.* (Winnipeg, 1976).

111. K. Yamaguchi, H. Matsumoto, N. Nakayama, and S. Ikegami, *Jap. J. Appl. Phys.* 15 (1976): 1575.

112. J. L. Shay, S. Wagner, K. J. Bachmann, E. Buehler, and H. M. Kasper, *Proc. 11th IEEE Photo. Spec. Conf.* (Scottsdale, AZ: IEEE, 1975), p. 503.

113. J. L. Shay, S. Wagner, M. Bettini, K. J. Bachmann, and E. Buehler, *IEEE Trans. Elec. Dev.* ED-24 (1977): 483.

114. R. A. Arndt, J. F. Allison, J. G. Haynos, and A. Meulenberg, *Proc. 11th IEEE Photo. Spec. Conf.* (Scottsdale, AZ: IEEE, 1975), p. 40.

115. J. M. Woodall and H. J. Hovel, *Appl. Phys. Lett.* 30 (1977): 492.

116. S. S. Chu, T. L. Chu, Y. T. Lee, *Proc. 14th IEEE Photo. Spec. Conf.* (San Diego: IEEE, 1980).

117. N. Nakayama, H. Matsumoto, K. Yamaguchi, S. Ikegami, and Y. Hioki, *Jap. J. Appl. Phys.* 15 (1976): 2281.

118. G. Bassak, *Electronics* (14 August 1980): 43.

119. R. F. Shaw and A. K. Ghosh, *Solar Cells* 1 (1980): 431.

120. H. Meier, "Application of the Semiconductor Properties of Dyes: Possibilities and Problems," *Topics in Current Chemistry, Physical and Chemical Applications of Dyestuffs* (New York: Springer-Verlag, 1976) pp. 85–131.

121. L. Lyons and O. Newman, *Australian J. Chem.* 24 (1973): 13.

122. A. K. Ghosh and T. Feng, *J. Appl. Phys.* 44 (1973): 2781.

123. A. K. Ghosh, D. Morel, T. Feng, R. Shaw, and C. Rowe, *J. Appl. Phys.* 45 (1974): 230.

124. P. J. Ruecroft, K. Takahashi, and H. Ullal, *Appl. Phys. Lett.* 25 (1974): 664.

125. P. J. Ruecroft, K. Takahashi, and H. Ullal, *J. Appl. Phys.* 46 (1975): 5218.

126. V. Y. Merritt and H. J. Hovel, *Appl. Phys. Lett.* 29 (1976): 414.

127. S. Kar, K. Rajeshwar, P. Singh, and J. DuBow, *Solar Energy* 23 (1979): 129.

128. M. S. Wrighton, *Technology Review* (May 1977): 30.

129. *Industry Week* (13 October 1980): 50.

130. A. Fujishima and K. Honda, *Nature* 238 (1972): 38.

131. M. S. Wrighton, A. B. Ellis, P. T. Wolczanski, D. L. Morse, H. B. Abrahamson, and D. S. Ginley, *J. Am. Chem. Soc.* 98 (1976): 2774.

132. D. E. Scaife, *Solar Energy* 25 (1980): 41.

133. N. N. Lichtin, *Chem. Tech.* (April 1980): 252.

134. H. P. Maruska and A. K. Ghosh, *Solar Energy* 20 (1978): 443.

135. A. Heller, K. C. Chang, and B. Miller, *Semiconductor Liquid-Junction Solar Cells, Proceedings*, vol. 77-3 (Princeton: Electrochemical Society, 1977), pp. 54–66.

136. K. C. Chang, A. Heller, B. Schwartz, S. Menezes, and B. Miller, *Science* 196 (1977): 1097.

137. J. Manassen, G. Hodes, and D. Cahen, *Semiconductor Liquid-Junction Solar Cells, Proceedings*, vol. 77-3 (Princeton: Elecrochemical Society, 1977), pp. 34–37.

138. Texas Instruments Corp., *First Quarter Report*, 1980, p. 14.

139. J. G. Posa, *Electronics* (6 November 1980): 39.

140. N. N. Lichtin, *Photogalvanic Processes, Solar Power and Fuels* (New York: Academic Press, 1977), pp. 119–142.

141. V. Srinivasan and E. Rabinowitch, *J. Chem. Phys.* 52 (1970): 1165.

142. D. E. Hall, J. A. Eckert, N. N. Lichtin, and P. D. Wildes, *J. Electrochem. Soc.* 123 (1976): 1705.

143. J. M. Mountz and H. Ti Tien, *Solar Energy* 21 (1978): 291.

144. R. W. Hosken, *Electro. Optical Sys. Design* (January 1975): 32.

145. M. Igbal, *Solar Energy* 23 (1979): 169.

146. M. Igbal, *Solar Energy* 22 (1979): 81.

147. D. M. Mosher, R. E. Boese, and R. J. Soukup, *Solar Energy* 19 (1977): 91.

148. T. M. Klucher, *Solar Energy* 23 (1979): 111.

149. T. A. Weiss and G. O. G. Löf, *Solar Energy* 24 (1980): 287.

150. L. W. James and R. L. Moon, *Proc. 11th IEEE Photo. Spec. Conf.* (Scottsdale, AZ: IEEE, 1975), p. 402.

151. D. L. Evans and L. W. Florschuetz, *Solar Energy* 20 (1978): 37.

152. H. J. Hovel, "Solar Cells," *Semiconductors and Semimetals*, vol. II, A. C. Beer and R. K. Willardson, eds. (New York: Academic Press, 1975).

153. L. W. James and R. L. Moon, *Appl. Phys. Lett.* 26 (1975): 467.

154. J. A. Castle, *Proc. 12th IEEE Photo. Spec. Conf.* (Baton Rouge, LA: IEEE, 1976), p. 751.

155. J. G. Fossum and E. L. Burgess, *Proc. 12th IEEE Photo. Spec. Conf.* (Baton Rouge, LA: IEEE, 1976), p. 737.

156. B. L. Sater and C. Goradia, *Proc. 11th IEEE Photo. Spec. Conf.* (Scottsdale, AZ: IEEE, 1975), p. 356.

157. C. Goradia, R. Ziegman, and B. L. Sater, *Proc. 12th IEEE Photo. Spec. Conf.* (Baton Rouge, LA: IEEE, 1976), p. 781.

158. T. Matsushita and T. Maimime, *Proc. IEEE Int. Electron Dev. Mtg.* (Washington, DC: IEEE, 1975), p. 353.

159. R. J. Schwartz and M. D. Lammert, *Proc. IEEE Int. Electron Dev. Mtg.* (Washington, DC: IEEE, 1975), p. 350.

160. R. M. Swanson and R. N. Bracewell, *Electric Power Research Institute* Report No. ER-478 (February 1977).

161. R. L. Bell, *Solar Energy* 23 (1979): 203.

162. F. DeMichelis and E. Minetti-Mezzetti, *Solar Cells* 1 (1980): 395.

163. G. Sassi, *Solar Energy* 24 (1980): 451.

164. E. C. Boes and B. D. Shafer, *Proc. DOE Ann. Photo. Program Rev. for Technology and Market Dev.* (Hyannis, MA: DOE, 1980), p. 46.

165. B. D. Shafer, E. C. Boes, and D. G. Shueler, *Proc. DOE Annual Photovoltaics Program Review for Technology and Market Development* (Hyannis, MA: DOE, 1980), pp. 29–59.

166. D. E. Carlson and C. W. Magee, *Appl. Phys. Lett.* 33 (1978): 81.

167. D. E. Carlson, *IEEE Trans. Electron Dev.* ED-24 (1977): 449.

168. E. D. Carlson and C. W. Magee, *2nd E.C. Photovoltaic Solar Energy Conf. Berlin* (Dardrecht, Holland: Reidel, 1979).

169. J. A. Grimshaw and W. G. Townsend, *Solar Cells* 2 (1980): 55.

170. A. G. Stanley, "Degradation of CdS Thin Film Solar Cells in Different Environments," *Technical Note 1970-33*. (Lexington, MA: Lincoln Laboratory, MIT, 1970).

171. A. Spakowski and A. Forestieri, *Proc. 7th IEEE Photo. Spec. Conf.* (Pasadena, CA: IEEE, 1968), p. 155.

172. D. Bernatowicz and H. Brandhorst, Jr., "The Degradation of Cu_2S-CdS Thin Film Solar Cells Under Simulated Orbital Conditions," *Proc. 8th IEEE Phot. Spec. Conf.* (Seattle, WA: IEEE, 1970), p. 24.

173. E. A. Demeo, Electric Power Research Institute Report No. ER-188 (February 1976).

174. F. Pfisterer, H. W. Schock, and W. H. Bloss, *Proc. 15th IEEE Photo. Spec. Conf.* (Baton Rouge, LA: IEEE, 1976), p. 502.

5

Projected Developments in Photovoltaic Solar Conversion Systems

The development of the solar cell as a major producer of electricity depends on not only its efficiency, stability, and cost, but also ancillary systems to convert the intermittent supply of direct current electricity to a steady, on-demand source of either dc or ac electricity of a voltage needed for a particular application. Certainly if the solar cell array can be used in conjunction with alternative sources of electricity, such as utility-supplied electricity or engine-powered generators, the main factor in increasing utilization is the cost of the array. For applications that are independent of such reliable alternative sources of electricity, the solar array must be supplemented with energy storage—such as a battery, fuel cell, flywheel, or pumped water stored for hydroelectric generation—so that electricity can be provided both when the sun is not shining and during peak power use when the solar array may not be large enough to meet power requirements.

Solar cells are commonly connected in either series or parallel arrays to provide power at voltages and currents that are acceptable for the load and suited to the means chosen to store excess power. A schematic diagram of a typical array of silicon solar cells arranged to drive a motor and charge a battery is shown in figure 5.1. By connecting 15 cells in series, each with a 0.45-V, 1.0-A output under full sun illumination, an output voltage of 6.75 V is reached. This is

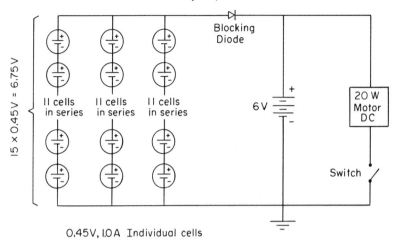

IA × 3 × 6.75V = 20.25 W Array output

15 × 0.45V = 6.75V

II cells in series   II cells in series   II cells in series

Blocking Diode

6 V

20 W Motor DC

Switch

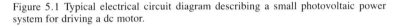

0.45V, 1.0A Individual cells

Figure 5.1 Typical electrical circuit diagram describing a small photovoltaic power system for driving a dc motor.

sufficient to charge the battery or run the motor if it is switched on. When insufficient sunshine is available, the output voltage drops below 6 V. Under these conditions, a blocking diode prevents battery discharge through the cells, and the battery provides power for driving the motor. The current from the array amounts to 3 A when the three series of 15 cells are connected in parallel. If the battery is sized appropriately, the system of electricity generation provides a constant source of about 20 W of power.

A schematic illustration of a more complicated system to provide alternating current is shown in figure 5.2. This system has an inverter that converts the direct photovoltaic current to alternating current with about 95% power efficiency, typical of modern inverters. Electromechanical or solid state inverters can be bought having capacity up to 1MW (megawatt). The solid state inverters are maintenance free and par-

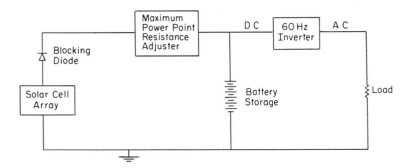

Figure 5.2 Schematic diagram of a photovoltaic power system for driving an ac load.

ticularly suited for residential use, typically in the 20-kW range. The system illustrated also involves a maximum power point tracker that adjusts the resistance across the solar array to maximize the power that can be withdrawn as, among other things, clouds, seasonal variations, and the position of the sun in the sky alter the solar input to the array. Although in a properly designed system the array will operate at maximum efficiency over a fairly wide range of solar input conditions, the maximum power point tracker allows for increased overall power output by ensuring that the voltage and amperage are at the knee of the amperage-versus-voltage curve (chapter 3, figures 3.8 and 3.9). In that case, for 100-mW/cm^2 illumination, V_{mp} and I_{mp} were 0.45 V and 24.4 mA, respectively. The load resistance should be 0.45/0.0244 or 18.4 Ω, from the $V = IR$ relation. When the insolation was reduced to half, the V_{mp} and I_{mp} were 0.43 V and 12.2 mA, respectively. Here the load resistance should be 0.43/0.0122 or 35.2 Ω. With a maximum power point tracker, the power output under the two conditions would be 11 and 5.2 mW. With a constant load of 18.4 Ω, the power output would be 11 and about 2.7 mW, respectively, because only a 0.225-V output would be reached at 0.0122-mA photocurrent across the 18.4-Ω resistance. Hence, the electronic arrangement to sense an array output and adjust resistance appropriately to provide maximum power output for any insolation level can be very helpful in reaching the full potential of high-quality photovoltaic conversion systems. The tracker consumes less power than it saves, although in large interconnected arrays for which cells are not matched in electrical properties, the load resistance is not as critical, and nearly optimal power output can be maintained with a single load resistance.[1] In the latter case, the investment in the power tracker might be avoided. But by the very nature of the electrical mismatches in the array, the output efficiency would be low.[2]

A broader conceptualization of a photovoltaic conversion system is illustrated in figure 5.3. In this case, the system may apply to a 20-kW system suitable for residential use, or it may apply as well to providing megawatts of power in a utility system. In a utility network, the type of storage may vary from nothing when the solar system is used to supplement a fossil fuel or nuclear-powered generation system, to pumped water storage when water is pumped upward and later released for hydroelectric generation. In a residential system, the storage may be a bank of batteries, fuel cells, or

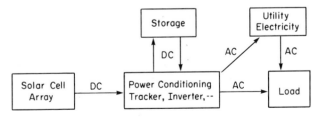

Figure 5.3 Schematic representation of a multifaceted photovoltaic power system involving power storage and alternative use of utility power for driving an ac load such as a residence.

a flywheel. If proper power conditioning is provided, residential systems may draw power from utilities when no sunlight is available and feed it back when an excess of photovoltaic power is generated. Some utility systems are experimenting with means of selling to and buying from their customers in this manner. Unfortunately, the daily peak load in power use typically lags the peak sunshine by 4–6 h, so the ability of the utility to buy solar-produced electricity from customers does not eliminate the need for having reserve capacity to meet peak demand. It does, however, provide additional power for driving pumped water storage, and it may also allow conservation of fossil fuel during the sunlit hours.

In an independent, general-purpose photovoltaic power system that provides power on demand, the main cost is for the solar cell array. The costs of arrays in 1980 were about $10,000/kW$_p$. The cost of lead acid battery storage was about $300/kW, but such a capacity may last only about 3 h before recharging is required. A 2-day storage to sustain a steady 1-kW power use will require 16 kW of capacity. Hence the actual cost would be about $4800/kW. The cost of an inverter and related power-conditioning equipment was about $120/kW for large systems[3] and between $400 and $1600/kW for a maximum power point tracking, utility interactive inverter for residential use.[4] The total system cost involving solar cells, storage, and power conditioning was about $15,000/kW$_p$. This did not compare well economically with a fossil fuel-powered plant at $500–1200/kW capital cost. However, the latter has ongoing costs of fuel supplies associated with it. Accounting for 1980 fuel costs, the figure of merit for comparison would be about $2000/kW for an oil-fueled electrical power plant.[5] Hence, the photovoltaic system is about eight times too expensive for general use in applications normally accommodated by utility electrical power. The rest of this chapter is devoted to projecting the probable future cost reductions in the major components of the photovoltaic power system.

Development of Solar Cell Arrays

As discussed above, the solar cell array is by far the most costly part of the photovoltaic energy conversion system. At $10/W_p$ for a 10% efficient array, the cost of a square foot is about $100. If on average in the United States 20 W/ft^2 were collected, then in 1 yr, 2 W·8766 h/yr, or 17.5 kWh, of electricity would be produced. Over a 20-yr life, this would amount to 350 kWh. Neglecting present value considerations of future foregone expenditures, we can simply divide $100 by 350 kWh and deduce that the array cost per kilowatt-hour of electricity is about 29¢. This is considerably higher than the average price of 4.5¢/kWh in the United States during 1980. For many remote areas, where utility power is unavailable, this may be a very reasonable price compared to use of primary batteries or fossil-fueled small generators that have to be maintained closely. However, for widespread use of solar cell arrays, the price must be reduced.

The various technical approaches that are being used to develop more efficient, less costly, and longer-lived solar cells were reviewed in the last chapter. Particularly for large-crystal silicon cells, where the technology for making 10–15% efficient arrays of 20-yr lifetime is well developed, cost reductions appear to be achievable by using less expensive grades of silicon, growing larger ingots, slicing more wafers per inch of ingot, and automating many of the manufacturing operations. Also, where silicon cells are used in conjunction with various sunlight concentration schemes, a large cost reduction appears feasible. Besides these engineering approaches to reduction of solar cell cost, various scientific approaches have taken shape in the last few years aimed at developing very thin, materials efficient cells that lend themselves to large-scale manufacture. Various "cadmium sulfide" cells, polycrystalline silicon, hydrogenated amorphous silicon, and gallium arsenide cells are being pursued in homojunction and heterojunction formats, including Schottky barrier, SIS, and MIS configurations. Photoelectrochemical cells capable of electrical storage are also being developed. Most of the emphasis in the research on small silicon crystals is devoted to achieving efficiency greater than 10% with a stable cell.

The US Department of Energy is currently taking a major role in the development of cost effective solar cell arrays. The development of solar cells was not appreciably supported by governmental research funds until the late 1970s. During the first half of the 1970s, the National Science Foundation provided some funding,[6] followed by increasing support by the Federal Energy Administration and the Energy Research and Development Administration, which set forth a national solar energy research, development, and demonstration program called the ERDA-49 program.[7] This program set a goal of reducing the cost of silicon solar cells from $20/W_p$ to $0.5/W_p$ by 1986—the goals being in terms of 1975 dollars. In the late 1970s, the Solar Energy Research Institute was organized under the Department of Energy to further support the effective development of photovoltaic conversion systems. Federal funding, though small compared to that for nuclear and synfuels development, has increased steadily in the past 5 yr, as shown in figure 5.4.

Considerable interest in the development of solar cells for commercial applications is being shown by US industrial

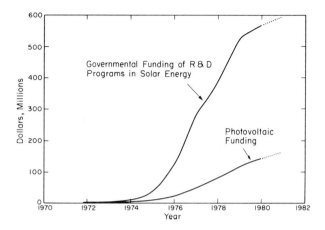

Figure 5.4 United States government funding of research and development in photovoltaic conversion systems.

corporations, particularly those already in the energy business. Many have substantial research programs. The Electric Power Research Institute, funded by many electric utility organizations, has an ongoing program to assess the potential use of photovoltaic conversion systems for providing commercial power. Various associations, professional societies, lobby groups, and journals in the area of solar cells have come into being in the last 5 yr. Seven companies are producing and marketing flat plate, large-crystal silicon cell arrays designed for terrestrial application.[8] About as many have bid on contracts to make concentration cells for Department of Energy demonstration projects.[9] All of this activity portends great effort devoted to improving solar cell arrays and reducing their cost. The cost goals set in the Department of Energy Photovoltaics Program expressed in 1980 dollars are shown in figure 5.5. The objective of this program is to reduce system costs to a competitive level in

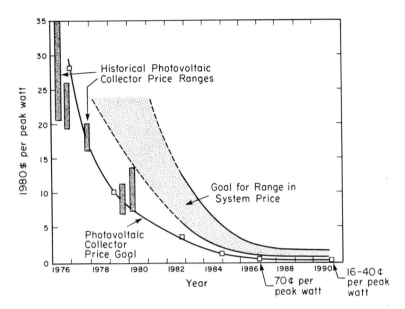

Figure 5.5 Photovoltaic array and installed power system goals in 1980 dollars as established in the United States Department of Energy Photovoltaics Program.

both distributed and centralized utility applications. The Research, Development, and Demonstration Act of 1978 authorizes a 10-yr, $1.5 billion program to achieve these and other related goals.[10] The goal of $0.16–0.40/$W_p$ for the solar cell array by 1990 is near that projected in technological studies of probable cost reductions in solar cells over the next 20 yr.[11,12] It is probable that the cost of photovoltaic arrays will decrease to 1/20th of their present cost over the next 15 yr.

Storage of Electrical Energy

Batteries

Batteries are classed as primary and secondary. Primary batteries are designed to be used once and discarded, whereas secondary batteries, such as those used in automobiles, can be recharged. The secondary battery is of interest for photovoltaic energy storage. Each cell of a battery consists of two electrodes, a separator, and electrolyte ions that complete the internal circuit and are involved in reactions at the anode and cathode. The separator keeps the electrodes apart so they do not short-circuit. Charge and mass transfer occur through the electrolyte, and chemical reactions occur at the electrodes because of differing electromotive forces between the active electrode materials. At the anode, an oxidative chemical reaction occurs that provides electrons and a negative potential on the electrode. At the cathode, a reductive chemical reaction occurs that uses electrons; that is, it provides a depletion of electrons and a positive potential on that electrode. Hence, when the two electrodes are connected through a load, a potential characteristic of the specific anode-electrolyte-cathode materials is established across the load, and a charge current will flow through the load. In secondary batteries, after the forward reactions have been expended in a discharge of the battery, the active electrode materials can be regenerated by an external source of reverse potential between the electrodes that exceeds the natural forward potential of the charged battery.

The electrical properties and storage capacities of batteries are determined by the type of electrode reactions, surface area of electrodes, and a myriad of material and design factors. The most commonly used and most promising batteries for lowest cost storage and performance improvements in the next 10 yr are the lead-acid, nickel-zinc, and nickel-iron batteries. Those that show great future promise but require considerably more technical and commercial development are the zinc-chlorine, lithium-metal sulfide, and sodium-sulfur batteries.[13] Nickel-cadmium cells are used where good portability is required, but not for general storage because of their relatively high price. Many others have been proposed and are in various phases of research. The main types of batteries considered for large power storage and some of their characteristics are listed in table 5.1. The exact characteristics of batteries of particular generic type vary greatly by design for various uses. For example, lead-acid batteries

Table 5.1 Secondary batteries in common use and under active development for photovoltaic energy storage and electric vehicles[a]

Common name	Electrochemical reaction	Open circuit voltage per cell (V)	Typical energy density (Wh/kg)	Recycles	Normal operating temperature (°C)
Lead-acid	$Pb + PbO_2 + 2H_2SO_4 \rightleftarrows$ $2PbSO_4 + 2H_2O$	2.0	30–100	200–2000	10–50
Nickel-zinc	$2NiO(OH) + 2H_2O + Zn \rightleftarrows$ $2Ni(OH)_2 + Zn(OH)_2$	1.7	55–90	300–600	10–60
Nickel-iron	$Fe + 2NiO(OH) + 2H_2O \rightleftarrows$ $Fe(OH)_2 + 2Ni(OH)_2$	1.4	45–70	500–2000	20–60
Zinc-chlorine	$Zn + Cl_2 \rightleftarrows ZnCl_2$	2.0	70–90	200–1000	20–50
Lithium-iron sulfide	$2Li + FeS \rightleftarrows$ $Li_2S + Fe$	1.6	60–120	200–1000	425–500
Sodium-sulfur	$2Na + 5S \rightleftarrows$ Na_2S_5	2.1	80–100	400–2000	300–350

[a]Data based upon information in references 13, 14, and 15.

designed for use in industrial trucks may have a 1500–2000-cycle life, whereas those for an automobile ignition system typically have a 200–400-cycle life. Batteries are typically specified in terms of the electrodes used, the energy density, the life in cycles, and cost. Because of the emphasis on potentially compact electrical storage for use in electric vehicles, development emphasis is being placed on the lithium-iron sulfide and sodium-sulfur batteries. They offer theoretical energy densities of 430 and 360 Wh/kg, respectively, versus 110 Wh/kg for the lead-acid (Pb-PbO_2 electodes) battery. However, they operate at high temperatures, which is a disadvantage.

The very high interest in electric vehicles and in peak power production portends intense research and development on increasing the life of the battery and decreasing its weight and cost. Various projections[16,17] indicate that there may be more than 30×10^6 electric vehicles in the United States by the year 2000. Also, the Department of Energy, the Electric Power Research Institute, and two power utilities are demonstrating a 6×10^6-lb lead-acid battery storage system capable of storing 30×10^6 kWh of electricity to meet peak demands for electricity.[18] Peak loads are typically met by using oil-fired turbines that have low capital costs of about \$125/kW but high operating costs because of low efficiency (24%) and high fuel costs. Base-load capital costs may be up to \$1200/kW, but the efficiency is 35–40%.[15] The oil cost per kilowatt-hour in the peaking turbine would be (36 \$/bbl·3412 Btu/kWh) ÷ (5,800,000 Btu/bbl·0.24), or 8.8¢/kWh. Lead-acid batteries cost about \$100/kWh single discharge. If the life is 1000 cycles, the average battery cost is 10¢/kWh. Of course, the battery must be charged with base-load-produced electricity. Because battery efficiency is typically 75% and base-load generation is 38% efficient, the overall efficiency is about 29%—higher than the 24% oil-fired peaking turbine.

The outlook for the secondary battery business is very good, and vigorous research and development efforts are expected to go along with a near doubling of the annual business from $7.2 to $14 billion during the 1980s. Because of manufacturing efficiencies associated with higher battery production and expected technological advances, the lead-acid battery cost per kilowatt-hour single discharge is expected[13] to decrease to $30–50. The costs for nickel-zinc and nickel-iron batteries are expected to decline in like degree. If the lifetime is extended to 2000 cycles (3 yr of daily cycling), this will yield an average cost of 2.5¢/kWh. During the 1990s, the average cost of electricity storage in residential and larger photovoltaic energy conversion systems should be below 2.5¢/kWh; the average US cost of electricity was 4.5¢/kWh in 1980.

Fuel Cells

Much like a battery, a fuel cell converts stored chemical energy into electrical energy. In the battery, electrode material is consumed upon discharge and regenerated during recharging. In a fuel cell, the electrode provides a catalytic surface on which the energy stored in reactants that are externally supplied to the cell can be released and converted to electricity. Fuel cells, in particular, the hydrogen-oxygen cell, offer the advantage of higher energy output per weight than batteries.[19] They are devices that generate electricity by constraining the oxidation of a fuel, hydrogen in this case, in such a way that electricity must flow through an external circuit for the reaction to occur.[20] The principles can be schematically shown in figure 5.6 for two types of cells. The hydrogen can be stored as a pressurized gas, as a metal hydride, or in various chemical forms that will easily release

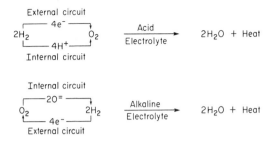

Figure 5.6 Schematic representations of principles of operation for two types of fuel cells.

it for use in the fuel cell. Oxygen can be derived from the air. The two fuels, H_2 and O_2, react indirectly to form water. The hydrogen is kept separate from the oxygen to prevent direct reaction, that is, burning. Because there is a driving force of 1.23 eV for the reaction of gaseous hydrogen and oxygen to form liquid water, if two catalytic electrodes (one for the oxidation of $2H_2$ to $4H^+$ and one for the reduction of O_2 to $2O^=$) are provided in a configuration such as in figure 5.7, the reaction can proceed with electrical current passing through an external circuit and ionic completion of the reaction to form water within the cell. The electrolyte used in the illustrated cell is an aqueous KOH solution. An acid electrolyte can be used as well. Platinum is an excellent but expensive electrode. Copper, silver, nickel (except in acid solutions), carbon, and nickel boride can be used as electrodes for the hydrogen oxidation. A porous carbon base with small amounts of platinum, silver, cobalt, copper, or nickel oxides on its surface is typically used for the oxygen reduction.[21]

In fuel cells the negative electrode or cathode, where a reactant is oxidized, is called the fuel electrode. In the above case the fuel is hydrogen. The positive electrode or anode, where a reactant is reduced, is called the oxidant electrode.

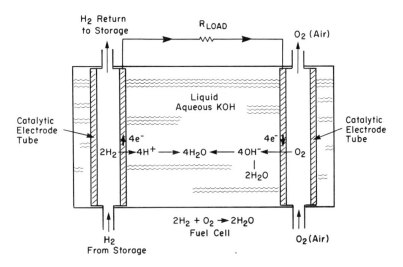

Figure 5.7 Fuel cell based upon the overall chemical reaction of hydrogen and oxygen constrained in a configuration to produce water internally and dc electrical power through an external circuit.

Oxygen is the oxidant in this cell and is reduced in the process of indirectly oxidizing hydrogen. The theoretical efficiency of this cell at ambient temperatures is 83%. The theoretical voltage across the cell is the difference between the standard electrode potentials in aqueous solution for the hydrogen and oxygen reactions (1.23 V). Other oxidants such as bromine, chlorine, or fluorine can be used, but with some peril to the catalytic electrode materials. For the hydrogen-bromine fuel cell incorporated into the photovoltaic converter shown in figure 4.18, the theoretical potential would be about 1.06 V. Other reactants such as methane, ammonia, hydrazine, and alcohol can be used in the fuel electrode. However, in photovoltaic systems, the electrolysis of water to produce hydrogen for storage and later use as a fuel in the fuel cells is a very simple and efficient process.

The efficiency of storage of photovoltaic power by electrolysis of water to produce hydrogen (90–95%)[22] and then using the hydrogen and air to generate electricity in a fuel cell (60–80%) is about 65%, whereas lead-acid batteries are about 70–75% efficient.[23] Fuel cell power is about twice as expensive and needs commercial development before it can come into general use. Fuel cells suffer from a short life, relatively high cost, and scarcity of catalytic metals.[24,25] With sufficient development, it is believed that electrolysis, H_2 storage, fuel cell regeneration of electricity, and inversion could be a practical method of storing energy for use in peak period generation of commercial electricity.[21,26] The H_2 would be stored in natural or man-made caverns and used upon demand. Thus, generating plants could be designed to operate with less reserve capacity, at a higher load factor, and at lower capital cost. If this proves practical for use with conventional plants, it should serve well in conjunction with photovoltaic conversion plants. Because fuel cells are not widely available as commercially developed sources of electricity, it is difficult to predict whether they will ever compete effectively with secondary batteries for storing photovoltaic power. They probably will not be economically competitive for the next 20 yr.

Pumped Water Storage

Hydroelectric power plants take advantage of the force of gravity acting upon water, which translates through a water turbine to a generator to produce electricity. By using excess electricity produced during off-peak hours to run a pump, water can be pumped upward to a higher potential energy in a reservoir. This can be released back through a water turbine to drive a generator to produce electricity. Reversible pump-turbine units and arrangements of turbines and high-pressure pumps on a common shaft can be used. Pumped water storage is presently a cost effective way to provide peak power electricity in many locations in the United States. In 1978 pumped water storage capacity exceeding 2×10^9 W was built.[27] The present installed capacity is more than 10^{10} W. This type of capacity costs $60–200/kW, depending upon the site and size of facility. Off-peak electric power from fossil-fueled or nuclear-powered base-load generators (35–40% efficient) is used to pump water into the hydroelectric storage reservoirs. This competes effectively with oil-fired peaking generators, which cost about $125/kW in capital cost plus the fuel to operate at a typical efficiency of 24%. The energetic efficiency of pumped water storage is about 75%, similar to battery efficiency.[15]

This type of storage technology is nearly fully developed and is cost effective when used in large power-generating systems, but not in the smaller community or individual residential power systems. It offers a very low cost way of storage when a photovoltaic generating station is used in conjunction with an existing hydroelectric system. When the demand for electricity exceeds the capacity of the existing hydroelectric system, a photovoltaic conversion system can be added to meet part of the demand in the sunlit hours. This in turn allows less release of water through the hydroelectric generator, that is, more storage of hydropower for later use in peak demand periods or nonsunlit periods. Typical daily demand for utility electricity and the insolation are as shown in figure 5.8.[28–31] This suggests that a portion of the potential hydropower could be stored during the midday hours, while the photovoltaic system provided power, and could be released in the late afternoon and evening hours to coincide with the peak demand for electricity. Furthermore, during July and August, when hydropower is typically least available, photovoltaic power would be most available. In essence, this is a significant application of photovoltaics that requires no additional expense for storage because the ex-

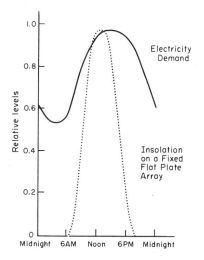

Figure 5.8 Relative demand for electricity in relation to typical solar incidence as a function of time.

isting hydroelectric plant would serve the purpose. Besides this naturally complementary relation of hydroelectric systems to photovoltaics, if pontoons are less expensive than neighboring land areas, the reservoir may provide an economical surface on which to place floating solar cell arrays.

Flywheels

A spinning mass in a low-friction mounting and environment can be a very compact means of storing energy. Friction losses can be minimized by using noncontacting magnetic bearings, high-density and high-strength spinning wheels that can operate at high speeds, and a low-pressure environment around the moving parts. Because the flywheel can have built into it a dc motor for driving it and an ac generator for providing output electricity, it is ideal as a storage medium between a photovoltaic array and an ac power load. In a flywheel-based system, the efficiency of storage and release of electrical power is about 68%.[32] This is nearly the same as a 65% efficiency expected in a maximum power tracker, battery, inverter system that accomplishes the comparable power storage and conditioning.

Flywheels are typically considered only for storage in relatively small applications, such as for residential photovoltaic conversion systems. Larger systems are cumbersome and difficult to design in reliable forms. With a size that stores about a 1-day electrical demand, the average total energy capture of a photovoltaic conversion system can be increased 46–58%.[33] A modern household requires a 10-kW peak capacity or greater, and a daily use of about 40 kWh should be assumed. In conjunction with a photovoltaic array, a flywheel with this capacity is estimated to add economic value particularly to photovoltaic energy conversion systems in remote locations where utility electricity is difficult to use.[34] Such a flywheel may weigh about 2 tons and operate at 15,000 rpm (revolutions per minute). Such flywheels are not readily available, but various manufacturers have estimated that they could be built for about $20,000 in quantities of 10,000 units per year.[32] This is $2000/kW$_p$, or $500/kWh single discharge. If the system has an 8-yr life of 365 cycles/yr with no maintenance, this will be 17¢/kWh. If volume increases, this cost may decrease; however, it is well above the present and expected cost associated with battery storage for such a system. A 10-kW present-day lead-acid battery system with a 4-h capacity (40 kWh) can be bought for about $4100. If the system lasts for 1000 cycles, this will be 10¢/kWh.

Photovoltaic Power Conditioning

Depending upon the end use, a small or large expense may be incurred to provide the proper power characteristics coming from the photovoltaic system. For example, a small transistor radio can be run by simply connecting a large capacitor across the output of the solar cell. This makes the system operate as a constant voltage source. Larger capacitors provide more protection against variations in solar incidence. For larger direct current applications, a battery can supplant the capacitor, and a blocking diode can prevent battery discharge through the solar array when the sun is not shining (see figure 5.1). For uses involving alternating current, an inverter is required. An inverter can be an arrangement of transistors, capacitors, and resistances, sometimes called a multivibrator or square wave oscillator, which "breaks up" the direct current into a square wave alternating current waveform. Silicon controlled rectifiers, bipolar transistors, and other solid state devices can be used. The inverter can be followed by a voltage multiplier, transformer, or an amplifier to meet the input requirements for a given use. If one wishes to maximize the average power output from the solar array, an electronic arrangement to detect the array output and adjust the resistance to maintain optimal power output at all solar incidence levels must be provided. In addition, when a photovoltaic system is used in concert with utility power, appropriate switching is required to provide power to a utility during sunlit hours and allow drawing from the utility when needed (see figure 5.3). Appropriate voltages, harmonic content, and phase locking must be provided to minimize distortion with the utility power grid. Also, if buy-back is a part of the business arrangements, appropriate metering of power fed into, and drawn out of, the utility grid is needed.

The power efficiency of inverters is 90–95%.[35] Because of the rapid development of solid state power-handling equipment, steady advances in efficiencies and cost reductions are expected. Maximum power trackers allow for a slightly higher overall power production from a photovoltaic system. Overall, the power conditioning in a complex system may trim the power output by up to 10%. The present cost of a 10-kW maximum power point tracking, inverting, power conditioner capable of interacting with a utility grid in a residential setting is $400–1600/kW. This is expected to be $190–320/kW by 1986.[4] Larger systems cost less per kilowatt.

Summary of Price Projections of Photovoltaic Conversion Systems

The use of photovoltaic power generation is tied inextricably to the costs associated with the solar cell array, power conditioning equipment, and supplementary storage systems. The costs associated with this unique method of electricity production have limited its application to areas in which alternatives are scarce, such as outer space and remote terrestrial applications. However, the cost of a photovoltaic system with moderate storage capacity has decreased from $200/$W_p$ in the 1960s to $10–20/$W_p$ today. The use of photovoltaic conversion systems in broader markets hinges on technological progress that will reduce the cost to the $1–2/$W_p$ range.

As discussed, considerable technological progress is expected in the three main cost components of a photovoltaic conversion system: solar cell arrays, energy storage systems, and power-conditioning equipment. Also, an increasing demand, almost doubling every year, will provide the incentive for automation in array production. Currently the cost of the solar cell array is decreasing more than 30%/yr.[36] This high rate of cost decrease is characteristic of high-technology industries in their infancy. My prediction of future photovoltaic converter costs, based upon the specific technological projections and programs cited in this chapter and also the synergism between developments in photovoltaic research and the electronics and secondary battery businesses, is presented in figure 5.9, given in terms of 1980 dollars. Those systems operated in the sunbelt with 3–5h of storage capacity and supplying alternating current would fall just above the dashed segment of the graph. The highest prices would be incurred in areas of the country with less insolation and in applications requiring more storage for stand-alone operation. In applications requiring no storage and little power conditioning, the price would be essentially that of

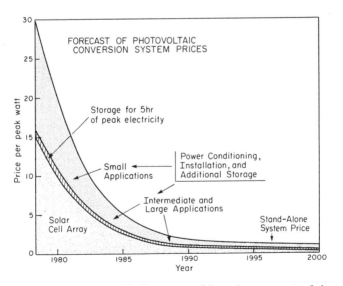

Figure 5.9 Projection of the future costs of the main components of photovoltaic power systems.

the solar cell array. This prediction includes a healthy measure of skepticism regarding the assumed rate at which technological possibilities can be incorporated into reliable production facilities. The forecast decrease in price of solar cell arrays is about 20%/yr until 1990, a slower rate than experienced in the last few years. The price should reach about 40¢/W_p by the year 2000. This would represent (in 1980 dollars) $4/ft² of 10% efficient arrays or $6/ft² of 15% efficient arrays.

References

1. G. M. Haas and S. Bloom, *Proc. 11th IEEE Photo. Spec. Conf.* (Scottsdale, AZ: IEEE, 1975), p. 256.

2. A. Luque and E. Lorenzo, *Solar Energy* 22 (1979): 187.

3. D. G. Schaeler and B. W. Marshall, *Proc. 12th IEEE Photo. Spec. Conf.* (Baton Rouge, LA: IEEE, 1976), p. 661.

4. C. H. Cox, *Proc. DOE Ann. Photo. Program Rev. for Tech. and Market Dev.* (Hyannis, MA: DOE, 1980), p. 281.

5. S. L. Leonard, *Proc. DOE Ann. Photo. Program Rev. for Tech. and Market Dev.* (Hyannis, MA: DOE, 1980), p. 465.

6. A. I. Rosenblatt, *Electronics* (4 April 1974): 99.

7. ERDA-49, *National Solar Energy Research, Development and Demonstration Program* (Washington, DC: US Government Printing Office, Superintendent of Documents, June 1975).

8. J. A. Eibling, *Solar Energy: An Assessment for Business*, B-TIP Rev. No. 2 (Columbus, OH: Battelle Memorial Institute, 1979), p. 18.

9. E. C. Boes and B. D. Shafer, *Proc. DOE Ann. Photo. Program Rev. for Tech. and Market Dev.* (Hyannis, MA: DOE, 1980), p. 29.

10. P. D. Maycock, *Proc. DOE Ann. Photo. Program Rev. for Tech. and Market Dev.* (Hyannis, MA: DOE, 1980), p. 2.

11. R. M. Moore, *Solar Energy* 18 (1976): 225.

12. J. A. Merrigan, *Sunlight to Electricity: Prospects for Solar Energy Conversion by Photovoltaics* (Cambridge, MA: MIT Press, 1975), p. 135.

13. W. J. Walsh, *Physics Today* (June 1980): 34.

14. D. L. Douglas and J. R. Birk, *Ann. Rev. Energy* 5 (1980): 61.

15. R. Whitaker and J. Birk, *EPRI Journal* 8 (October 1976): 6.

16. B. Horovitz, *Industry Week* (19 September 1980): 80.

17. Gulf and Western, *Wall Street Journal* (6 June 1980): 11.

18. R/D News, *Industrial Research and Development* (January 1981): 43.

19. K. R. Williams, *An Introduction to Fuel Cells* (New York: Elsevier, 1966), p. 134.

20. S. W. Angrist, *Direct Energy Conversion* (Boston: Allyn and Bacon, 1965), p. 326.

21. A. McDougall, *Fuel Cells* (New York: John Wiley and Sons, 1976).

22. W. E. Morrow, Jr., *Tech. Rev.* 76 (1973): 31.

23. F. Daniels, *Solar Energy* 6 (1962): 78.

24. R. Boll and R. Bhada, *Energy Conversion* 8 (1968): 3.

25. B. Baker, *Fuel Cell Systems—II, Advances in Chemistry Series*, R. F. Gould, ed. (Washington, DC: American Chemical Society Publications, 1969).

26. A. Bruckner, II, W. Fabrycky, and J. Shamblin, *IEEE Spectrum* (April 1968): 101.

27. R. C. Dorf, *Energy, Resources, and Policy* (Reading, MA: Addison-Wesley, 1978), p. 149.

28. M. A. Maidique and B. Woo, *Tech. Rev.* (May, 1980): 25.

29. W. D. Marsh, *Requirements Assessment of Photovoltaic Power Plants in Electric Utility Systems* (Palo Alto, CA: Electric Power Research Institute, June 1978) EPRI ER-685-54, Vol. 1, Project 651-1, Summary Report.

30. M. L. Brown, *Proc. DOE Ann. Photo. Program Rev. for Tech. and Market Dev.* (Hyannis, MA: DOE, 1980), p. 477.

31. W. Dickter, *Proc. DOE Ann. Photo. Program Rev. for Tech. and Market Dev.* (Hyannis, MA: DOE, 1980), p. 407.

32. P. O. Jarvinen, *Proc. DOE Ann. Photo. Program Rev. for Tech. and Market Dev.* (Hyannis, MA: DOE, 1980), p. 205.

33. General Electric Space Division, *Applied Research on Energy Storage and Conversion for Photovoltaic andWind Energy Systems*, Final Report, January 1978.

34. T. L. Dinwoodie, *Flywheel Storage for Photovoltaics: An Economic Evaluation of Two Applications* (Cambridge, MA: MIT Energy Laboratory, February 1980).

35. D. R. Smith, *Proc. DOE Ann. Photo. Program Rev. for Tech. and Market Dev.* (Hyannis, MA: DOE, 1980), p. 271.

36. D. Costello and P. Rappaport, *Ann. Rev. Energy* 5 (1980): 343.

6

Business Opportunities in Photovoltaic Energy Conversion Systems

Because the manufacture of photovoltaic systems is only in its infancy, it is typically accomplished by job shop or cottage industry methods. Future business opportunities depend on incentives to technical entrepreneurs for the exploration of economical processes for mass producing reliable, long-lived solar cell arrays of lower cost and ever increasing conversion efficiencies. The present state of the art was achieved by a relatively small group of researchers and entrepreneurs in individual efforts. Some incentive for production of high-efficiency stable silicon cells has been provided by space exploration efforts. However, because the alternative sources of energy in space are minimal and costly and this market is small, large manufacturing capacity, which would yield lower array costs, has not been needed.

It has been only during the last 5 yr that increasingly significant markets for terrestrial photovoltaics have been addressed. This was stimulated by concerns about potential energy shortages in the United States and by the constantly increasing prices of fossil-fueled electrical power generation. An increasing support for research and development of photovoltaic conversion systems by governmental funding (see figure 5.4) stimulated a broader interest in industry and also more private investment in the development of less expensive systems offering business opportunities in large markets. One might classify present R&D efforts into four categories, involving academic institutions, the present suppliers of solar cells, electronics companies, and energy companies. Since the entry by, among others, RCA, Texas Instruments, Sanyo, Motorola, IBM, Westinghouse, General Electric, Mobil, Exxon, ARCO, and Shell Oil, R&D in this area has become a substantial and broadly balanced effort aimed at developing photovoltaic energy conversion into a practical technology for terrestrial production of electricity.

Key Factors in the Use and Development of Photovoltaics

Attention to the development of photovoltaic technology by more people skilled in the techniques of design and production may be expected to reduce costs. Mass production techniques will introduce economies of scale at the production level. But mass production requires mass markets, and mass markets are not easily developed in fields such as this, in which other inexpensive and convenient sources of energy are already available. Various factors besides price will influence the development of markets for photovoltaics. Who wants to have electricity at the whim of the weather? Who wants to look at a lot of black solar collectors? Who wants to be responsible for upkeep of a system? Who wants, or has the money, to make now that large capital investment in return for "free" energy in the future? Who wants to be in the vanguard? Who desires to be energy independent? The psychology of the marketplace and the financing options available certainly will influence the future commercialization of photovoltaic energy converters. If conventional sources of electricity are unavailable, as in remote locations, or become unreliable relative to individually owned supplying devices, people will pay an appropriate price for the relative security of their own supplies.

Photovoltaic power generation uses no moving parts, no liquid or gaseous wastes are generated, no unusual safety problems arise, no radioactivity is released, and for most uses, no thermal problems are encountered other than those normally associated with a dark roof. Space used for solar collection can be used for other purposes as well, such as for roofs and walls. The only obvious associated environmental problems are the visible undesirability of dark surfaces, any problems of safety in manufacturing associated with handling, for example, silicon, CdS, and GaAs, and those problems associated with battery, fuel cell, or other means of storage of energy.[1] Cadmium sulfide and GaAs present some toxic hazard if burned, but such small amounts are used that this should be manageable in a noncongested setting.

Because the "fuel" for a solar energy converter is free, the variable cost of power output is nearly independent of the conversion efficiency. The most significant economic factor is output per unit capital cost properly adjusted for the time value of money, depreciation, taxes, and relatively minor maintenance costs. Because capital depreciation continues regardless of the use rate of the solar converter, the load factor (percentage of time that full capacity is used) is very significant. Terrestrial solar energy is a doubly periodic function with periods of 24 h and 365 days onto which is imposed the random fluctuations of cloud cover. The maximum load factor for photovoltaic power systems would be about 40% in places like Phoenix and Las Vegas (See figure 2.4). The average load factor over the United States would be about 17%, that is, 17 W/ft^2 ÷ 100 W/ft^2, the average solar incidence divided by the incidence from direct sunlight on a plane normal to the sun. The cost per kilowatt-hour output goes down as the load factor goes up. Hence, the sunbelt areas will be first to find the economics favorable for using photovoltaics. If all other factors such as degradation rate and maintenance are equal across the United States, terrestrially generated photoelectricity in Phoenix would cost about 42% as much as the average across the nation, that is, 17/40, the average load factor divided by the Phoenix load factor.

Certainly the cost of photovoltaic arrays should decrease as production quantities increase. However, except for quantity discounts, the individual investor would not realize economies of scale in building larger arrays. The output of a collector is directly proportional to its area. Hence, more capacity would be gained simply by adding more modules of the same solar cell arrays. This is in sharp contrast to most other forms of electricity generation. Thermoelectric and hydroelectric generating stations provide less expensive power as the size increases. Diesel-generated electricity used, for example, in remote locations or for special equipment like locomotives is much less expensive, the higher the capacity. This is made possible by the capital cost savings as unit size increases. Because small-scale photovoltaic conversion is as cost effective as large-scale, the cost of distribution of power from central locations can be avoided by providing photovoltaic conversion at the site of electricity use. The diffuse nature of solar energy lends itself to use in sparsely populated and remote areas.

These attributes of solar energy and photovoltaic energy conversion may be considered advantageous compared to those of conventional sources of electricity involving massive investment; visual undesirability of power lines; depletion of fossil fuels that could go into petrochemicals, fertilizers, and plastics; and the environmental effects of burning. The latter are predominantly social expenses, however, which cannot be fully assessed by a homeowner when deciding whether to connect into a commercial power system, which provides a reasonably secure supply of electricity at about \$400/yr, or to construct his own supply at an immediate cost of several thousand dollars and future responsibility for maintenance and operation.

The demand for electricity is a derived demand. Consumers buy it to satisfy other demands. Electricity per se is rarely a concern to the end users. Therefore, they will make decisions about electricity mostly on the basis of cost, availability, convenience, reliability, and the timing of monetary outlays. To compete favorably with fossil fuel generation for the largest markets, photovoltaic generation must offer electricity at a comparable price with provision for continuous availability and equal reliability. With proper financial and ownership arrangements, the monetary flows can be arranged to follow patterns suitable to the consumer, although if the consumer owns his own photovoltaic facility, the initial outlay (or the realization of what the long-term outlay is) may be a formidable barrier to marketing of photovoltaic converters—even if the produced electricity is as inexpensive as buying electricity from conventional utilities.

Figure 6.1 is a diagram of the key factors that govern the development of photovoltaic electricity production and the demand for it. The demand for electricity generated by photovoltaic energy conversion depends on the factors pointing toward it in the diagram: the overall demand for electricity; the reliability, price, and pollution associated with conventional generation; and the reliability and price associated with photovoltaic power generation. In figure 6.1, an increase in the level of a causing factor at the tail of an arrow results in an increase (+) or decrease (−) in the level of the effect at the point of the arrow. If the reliability of conventionally generated electricity goes down or the price or pollution goes up, the demand for photovoltaic conversion goes up. The higher the reliability of photovoltaic power, the higher the demand for it. The lower its price, the higher the

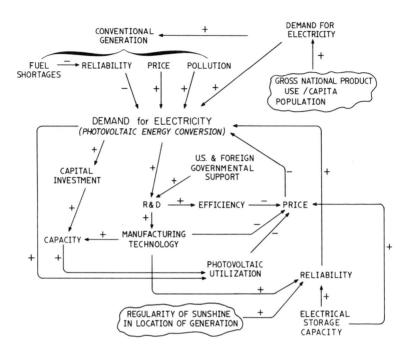

Figure 6.1 Schematic representation of the factors that affect the development of photovoltaic converters. An increase in the level of a cause (tail of arrow) results in an increase (+) or a decrease (−) in the level of an effect (point of arrow).

demand. With stimulation of research and development, increased solar cell efficiency and reduction in price of photoelectricity ensues, which yields higher demand. Higher demand stimulates investment, ultimately resulting in increased use and lower price. Similarly, increased insolation makes the reliability greater, so that demand is greater in sunbelt areas. An increase in electrical storage capacity increases the reliability as well as the price of photovoltaic power generation. The optimal balance between storage capacity, price, and reliability that is acceptable for a given application determines the level of storage incorporated into a system.

If the technological projection shown in Figure 5.9 becomes a reality, the demand for photovoltaic conversion systems should increase monumentally. The key question is, Will the volume demanded at the $10/W_p$ and $5/W_p$ price levels generate enough business to fund the continuing R&D and capital investment necessary to drive the costs continually lower and finally reach the $1–2/W_p$ values that allow truly low-cost mass production for supplying the utility companies with generating capacity? Answering this question will involve the consideration of the market.

Markets for Photovoltaic Energy Conversion

Nearly all of the electricity used in the United States is supplied by large electrical utilities. Exceptions occur where portability or remote use is required. Primary batteries (those used once and discarded) such as the carbon-zinc LeClanche "flashlight," alkaline-manganese, and mercury cells supply the electricity for powering, among other things, flashlights, toys, camera photoelectric eyes, transistor radios, and clocks. This is a $900 million business in the United States and more than $2.5 billion worldwide. Secondary batteries (rechargeable) such as lead-acid, nickel-cadmium, silver-zinc, and silver-cadmium cells supply the electricity for starting, lighting, and ignition in automobiles, boats, and aircraft and powering electric toothbrushes, shavers, fork lift trucks, golf carts. This is a $3.4 billion business in the United States and more than $8 billion worldwide. Secondary batteries are generally recharged by electrical power from electric utility companies or from engine-powered generators. Both the primary and secondary battery businesses are expected to double in the next 10 yr. Still other electricity is generated by generators driven by diesel or gasoline engines, particularly of the industrial size used at remote construction sites.

The present annual use of utility-provided electricity in the United States is about 2.3×10^{12} kWh. This is a $100 billion business. It is expected to grow in volume at about 3.5%/yr through the end of this century. The growth in electricity use is due in large part to its cleanness and convenience. Convenience, reliability, and cleanness are certainly of prime importance to a customer. Electricity offers energy on demand by the flip of a switch. Photovoltaic energy conversion is an alternative method for supplying electricity. Hence, it will be used predominantly as a substitute method for producing electricity in existing markets and must compete by virtue of superior characteristics such as lower price, greater reliability, and better availability.

Because of its price, photovoltaic energy conversion cannot compete with fossil-fueled electric power generation in the well-developed part of the world. However, photovoltaics can be, and are being, used in various smaller segments of the overall market,[2] notably in space, offshore oil drilling stations, buoys, forest service watchtowers, electronic watches, radio repeater stations, and remote communications networks. Solar cell arrays are penetrating the recreation market as battery chargers on boats and golf carts, at campsites, cabins and lodges, and in various remote locations where small generators driven by internal combustion engines are the only sources of electricity. They are used in pumping water for irrigation of remote farmland, in providing power at construction sites before utility power is available, and in powering roadside flashers. They are used also in reducing electrolytic corrosion in cases where two dissimilar metals are in contact in an environment that is electrolytic. This is called cathodic protection; essentially, a reverse electrical bias is put across the "battery" formed by the metals. Bridges and pipelines are prime candidates for this type of protection.

The market for photovoltaic energy conversion systems can be subdivided in order to get a better insight into the ways in which the market can be expected to develop as the price of photovoltaic electricity declines. Several studies have made a subdivision on the basis of near-term, intermediate, and long-term markets.[3-12] The rationale and assumptions for market penetration differ among these studies, but they offer studied projections of annual sales as a function of the price of photovoltaic systems.[3-12] The various forecasts,[3-12] expressed in terms of 1980 dollars, are presented in table 6.1

The existence of sufficiently broadening markets as the price decreases is very important. As new technical breakthroughs make lower prices possible, a new market must be

Table 6.1 Market forecast ranges for photovoltaic energy conversion systems[a]

| | Potential annual sales | | | |
	Price per peak watt (1980 dollars)	Megawatts (peak)	Dollars (millions)	10% efficiency arrays (million ft²)
Near-term	14	0.4–10	5.6–140	0.04–1
Intermediate-term (8 yr)	4.2	2.6–75	10.9–315	0.26–7.5
	1.4	30–500	42–700	3–50
Long-term (15 yr)	0.7	100–2000	70–1400	10–200
	0.14–0.42	5000–100,000	70–42,000	500–10,000

[a]See references 3–12; forecasts made prior to doubling of oil price in 1979–1980.

available that is sufficiently large to support the capital investment for larger and advanced production capabilities. These markets allow the prices to continue to decrease as production volume increases. Penetration into the utility power generation market, the largest market for electricity, is not probable without the existence of sufficiently large markets willing to pay premium prices for electricity, that is, those presently served by various batteries and engine-powered generators or not served at all.

Because of the increasing returns to scale in conventional generation, electricity is typically least expensive where large quantities are required. The price per kilowatt-hour is least in the large industrial applications and higher in the residential sector, where a household typically requires only about 8900 kWh/yr. In smaller applications, where utility power is not available, electricity may cost several times to several thousand times the cost of residential power. For example, the average cost of electricity in the United States in 1980 was 4.5¢/kWh. The cost for generating power with a typical portable engine-powered generator is 15–80¢/kWh, depending on the generator size, price, efficiency, cost of fuel, and so on. A secondary lead-acid battery yields electricity for 15–95¢/kWh depending on the battery's recycle life, size, price, recharging cost, and so on. A nickel-cadmium battery for use in a rechargeable electric shaver or toy typically yields electricity at $1–10/kWh, and a primary battery for use in an electric watch or calculator may yield electricity for $500–4000/kWh.

Where only a small amount of power is required, very high prices per unit are acceptable. Hence, when we consider near-term to intermediate-term to long-term markets for photovoltaics, we are dealing with gradually decreasing price structures and markets involving, respectively, small (watt), intermediate (kilowatt), and large (megawatt) applications. The inherent modularity of photovoltaic arrays makes it possible to apply a common technical design to these various applications—an ideal feature. Use of the same module to serve many such applications brings with it the potential for economies associated with mass production, which will manifest themselves as lower prices in all segments of the market. Thus, if the price should reach a level that results in, say, 5×10^6 ft^2/yr production to supplement the peak power generation capacity of a particular utility, the economies of scale in production may reduce the array price and broaden its penetration in the irrigation pumping or individual residential powering market segments. This in turn would tend to provide more incentive for manufacturing upgrading and expansion, lowering of price, and further penetration into the utility generation market. This cycle is possible as long as increasing manufacturing efficiencies are available at larger volumes. There is considerable debate in the technical community whether an artificial demand generated by the government could trigger such a cycle so that the prices of silicon solar cell arrays would decrease to the point where a continued "real" demand would provide the basis for a new photovoltaic conversion industry.

Addressing the markets for photovoltaic conversion systems at today's prices, various studies indicate potential annual sales of 3–7 MW$_p$.[4,11,12] At a system price of $10–20/W$_p$, this amounts to a potential $30–140 million annual business. Although market information is sketchy, various estimates indicate that the sales volume of photovoltaic arrays has grown by over 60% annually as the price decreased by slightly over 30% annually for the past few years.[4] The 1980 production of photovoltaic arrays was nearly 3 MW$_p$, with complete system sales of $50–70 million. The systems were sold chiefly for remote and small power applications, as mentioned earlier in this section. Basically they are being substituted for primary batteries that deliver energy at an average cost of about $30/kWh, secondary batteries that deliver power at $100/kWh single discharge but can be recharged 100–500 times (that is, $0.20–1/kWh), and various gasoline- and diesel-powered generators that yield electricity at 20–90¢/kWh. The cost of photovoltaic electricity per kilowatt-hour varies by application but can be estimated as follows. A square-foot array in an average US location should yield about 20 W average insolation times 0.10 efficiency or 2 W average power. If this were delivered for 10–20 yr, that is, 87,660–175,320 h, then a total of 175.32–350.64 kWh would be produced. An installed square-foot array costs about $150, that is, $15/W_p times 10 W_p/ft². Division by the number of kilowatt-hours produced over the array lifetime yields a cost of 43–86¢/kWh. As a point of comparison, if the price of the photovoltaic array were $1.5/W_p (the price projected for about 1988; see figure 5.9), then the cost would be 4.3–8.6¢/kWh, which is the typical cost of electricity supplied by utilities.

The intermediate markets where photovoltaic power systems should be competitive at a price of about $5/W_p are expected to be water pumping for irrigation and potable water use, general power supply for remote homes, villages, and communities, and street and highway lighting. Many users in these markets pay about 25–35¢/kWh for electricity. The world intermediate market is estimated to be more than three times that in the United States because the utility grid systems are not as highly developed in many nations. Water pumping is one of the most probable markets amenable to development. Demand for water is typically coincident with sunshine, so that storage capacity should be minimal, but where necessary, the water, rather than the electricity, can be stored. Units in the 0.1–1.5-kW range (0.13–2 hp) are prime candidates because they can be developed by using a modest-sized array (10 by 15 ft or less). Because the pumping unit is small, competitive gasoline- or diesel-powered pumps are relatively expensive and, of course, have to be regularly supplied with fuel—a key detriment in remote locations. Windmills for pumping water cost about $2000/hp in these sizes, and when the sunshine is most direct and water is most needed, wind is typically least available. The world potential for irrigation pumping has been estimated at up to 23,500 MW_p by 1985—a huge market.[13] If only 1% of this market were penetrated each year, a 235-MW annual business would be generated. At $5/$W_p$, this would be $1.2 billion in annual sales.

Another potentially large intermediate-term application for photovoltaics is in the supply of electricity to remote military installations.[14] This has been estimated by the Department of Defense at up to 100 MW. Even though the two intermediate market segments discussed here may amount to a demand of 200–300 MW/yr, the largest segment and most unquantifiable is that which will serve the parts of the world where electrical distribution grids are not readily available. The modularity of photovoltaic arrays will allow investment in a small system to provide the most essential lighting, refrigeration, and communications needs, while allowing the potential for easy addition of modules to provide for increasing demands as the price of modules decreases and ability to buy increases.

The intermediate market size has been estimated at 50–500 MW annually.[4,11,12] By comparing the cost of electricity during the rural electrification of the United States in the 1940s (20–30¢/kWh in 1980 dollars) with the cost of electricity from a $5/$W_p$ photovoltaic array (about 15–30¢/kWh), it is easy to visualize a very strong developing market for photovoltaics if effective marketing and distribution channels can be established in the developing nations. Considerable interest in the potential of the international market has led to US plans for demonstrating applications of photovoltaic energy systems and facilitating their widespread use in other nations.[15] Table 6.2 presents a Sciences Applications Inc. assessment of the world sales in 1978 and 1979. This study was made for the DOE Solar Energy Research Institute.[16] About half of the sales were exported from the countries of manufacturing origin. The main markets outside of the United States, European, and Japanese markets appear to be Greece, Israel, Australia, and the Middle East and West African nations. Photovoltaic-powered water pumping systems are being built for use in Senegal, Nigeria, Mali, Cameroon, Rwanda, and Upper Volta. Most of today's markets in these areas may be classified as product demonstration markets and are likely to be very critical growth markets as the price of photovoltaic power decreases to $1.5-6.0/W_p$. Development of this market may take several years, but as it occurs, the estimate of a 50–500-MW/yr intermediate market for photovoltaic system sales may be an underestimate.

The long-term and largest market for photovoltaics is in general electricity supply in concert with fossil- and nuclear-

Table 6.2 Overview of world sales of photovoltaic cells and arrays[a]

	Sales in megawatts peak power			
Countries of origin	1978		1979	
United States				
Domestic	0.38 ⎫	76%	0.65 ⎫	77%
Exported	0.30 ⎭		0.58 ⎭	
European (Predominantly Exported)	0.175	19%	0.30	19%
Japan	0.045	5%	0.07	4%
Total	0.90	100%	1.60	100%

[a]Data derived from reference 16.

fueled utility power grid systems. This is expected to occur increasingly as the price of photovoltaic conversion systems decreases across the range of $2–0.5/$W$_p$. Market penetration will occur by substitution for utility power in locations where it is most expensive; by use within the utility grid system, where economical, for supplementing the power output in peak demand periods; and ultimately by full integration into the utility power system as the cost and reliability of electricity from the photovoltaic power system becomes competitive with that from fossil- and nuclear-fueled generators. Various studies indicate that relatively small single-residence systems (3–20 kW$_p$) will be economically practical when the price reaches $2.6–1/$W$_p$.[17–21] Larger power systems of 500–1000-kW$_p$ sizes for commercial, industrial, and institutional applications will be feasible in about the same price range, and the larger 50–100-MW$_p$ utility power systems will become cost competitive at the low end of the range, depending on geographic location and primary fuel prices.[22–24]

The present price of electricity from a stand-alone residential photovoltaic power system is too high to compete with utility-supplied electricity, unless the residence is so remote that miles of power lines would be needed to serve the residence. For example, if a 15-kW service were desired (136 A in a 110-V system) with photovoltaic capacity during sunshine hours to provide peak capacity without storage, a 1500-ft^2 array with 10% efficiency would be needed:
1500 ft^2 · 0.10 efficiency · 100 W/ft^2 peak insolation = 15,000 W$_p$.
This would cost about $8/$W$_p$ or $120,000. If the household used the US average annual amount of residential electricity, 8900 kWh, this would be 24.4 kWh/day. If a 2-day lead-acid battery storage system were bought at the prevailing price of $100/kWh single discharge, it would cost about 48.8 kWh

· $100/kWh, or $4,880. The power conditioning system, including an inverter to supply ac power, but not a maximum power point tracker, would cost about 15 kW · $300/kW = $4500. If we assume the residence is new and the installation costs are offset by savings of normal roof material where the arrays are situated, then the cost would be array, $120,000; battery, $4,800; power conditioning, $4,500; total, $129,300.

Simple arithmetic calculations indicate this to be a cost of 97¢/kWh over a 15-yr system life. Given that the expenditure is now and savings on electricity costs are over a future 15 yr, a present-value analysis yields an even higher average cost per kilowatt-hour, assuming future yearly increases in electricity costs do not exceed the annual return on other investments one could make with the money. The typical cost of utility-provided residential electricity today is 4–6¢/ kWh. By far the most costly factor in the residential photovoltaic power system is the array. This is where major future cost reductions are anticipated. As the array price decreases from $8/$W_p$ to 50¢/W_p in the 1990s, the residential application will become more attractive, particularly where integration into the utility system reduces the size of the array and the storage needed to meet the household's peak electricity need and stability of supply. If one of every ten new homes were built with 1000 ft^2 of photovoltaic arrays as roof material, there would be an annual market of 0.1–0.2 million homes times 1000 ft^2 (10 kW$_p$), or 100–200 million ft^2 (1–2 billion W$_p$). This does not include retrofitting any of the present 80 million US households or the market for replacement of units.

It is unlikely that individual residential photovoltaic power systems will ever account for more than 10% of electrical power in the residential sector in the United States. At the photovoltaic array price levels that would make such a market penetration economically attractive, utility companies would also find photovoltaics attractive for providing supplementary power in the distributed grid system. Hence, as the cost of solar cell arrays is reduced, before most householders find it economically attractive to have their own system versus buying power from a utility, the utility will incorporate photovoltaics into its generation system to minimize its own costs. In countries where electrical power distribution systems are well developed and extensively available (that is, the initial capital costs for distribution have already been paid), the utility power companies, by using their choice of primary energy sources (including solar cells), should be able to supply electricity to households at a price that is as low as can be realized by the average consumer with an individual system. This would not be the case in the less developed areas of the world, where efficient, cost effective utility grid systems have not reached villages and rural residences.

Penetration of the utility and large industrial generation systems represents the largest market for photovoltaic arrays in the next 50 yr. Many studies of the potential for serving this market suggest that when the price of electricity from photovoltaic arrays reaches $1000–2500/kW_p$, depending on solar incidence and primary fuel costs, penetration of this market will begin.[22–27] The central utility power systems in the United States alone have a capacity of about 600,000 MW. About 35–40% of this is reserve capacity that can be brought on line during peak demand periods, such as mid-afternoon on the warmest days in July and August when industrial, commercial, and residential power demand is greatest.[28] Utilities typically operate at less than a 65% peak load factor using steam-driven generators, with quickly responsive gas turbine and diesel-powered generators available for meeting the peak demands. The capital cost of the latter peaking generators is less than that for base-load steam-driven generators ($125–300 versus $800–1200/kW), but they are less efficient in converting primary fuel energy to secondary electrical energy (24 versus 38%). Coal and nuclear plants, because of the associated large annual capital depreciation costs, have the highest operating costs among conventional generating options when operated for peaking capacity (0–1200 h/yr). However, because the fuel costs are lower than for oil- or gas-fired generators, coal and nuclear plants offer the lowest operating costs when operated for base-load capacity (3500–6000 h/yr). Oil- and gas-fired generators have the lowest operating costs for peaking because of the lower capital depreciation costs associated with them. However, because of the rapid rise in primary fuel costs, fuel costs rather than capital costs are increasingly the determining factors of the power to be used in intermediate capacities (1200–3500 h/yr). If an oil-fired generator operates at 24% efficiency, the cost of fuel to provide a kilowatt-hour of electricity is 6.2×10^{-4}¢/Btu · 3412 Btu/kWh ÷ 0.24, or 8.8¢. If a coal-fired base-load generator operates at 38% efficiency, the cost of fuel per kilowatt-hour of electricity is 1.6×10^{-4}¢/Btu · 3412 Btu/kWh ÷ 0.38, or 1.5¢.

In the US central thermoelectric generating system, with its particular mix of new and old capital equipment, the present total cost of peaking power is about 12–18¢/kWh, and that of base-load power is 3–7¢/kWh, depending on the particular fuel used and the generating equipment that is "in place" in a particular area of the country. It is necessary, then, for photovoltaic conversion systems to provide electricity for about 15¢/kWh to begin to penetrate the market for peaking capacity. Using unsophisticated calculations, one can approximate the maximum price one would pay for the photovoltaic system. If a square-foot array tilted toward the sun intercepted an average of 20 W of sunshine over a year and converted it to electricity at 10% efficiency, a total of 20 W·0.1·8766 h/yr, or 17.5 kWh, would be produced. A square-foot array intercepts about 100 W of direct sunlight shining normal to it, so that the peak power rating would be 10 W_p. The yearly value of the electricity from this square-foot, 10-W_p source would be 17.5 kWh·15¢/kWh, or $2.62. If the value of electricity stayed constant and this 10-W_p array lasted 20 yr, it would produce electricity worth $52.40. This is a value of $5.24/$W_p$, or $5240/k$W_p$. Because of some maintenance and replacement costs, as well as the risk that the value of electricity produced in the future may not increase as rapidly as the value of other investments, the photovoltaic array would actually command a price somewhat lower than $5240/k$W_p$.

More sophisticated analyses have been made to establish the price at which photovoltaic power converters will become feasible in the central power station market[19,25], but most were made obsolete by the doubling of the price of oil in 1979–1980. Recent publications indicate that public utilities, particularly those that use oil for generation, would pay about $1800–2200/k$W_p$, while investor-owned utilities would pay about $1400–1700/k$W_p$ for photovoltaic conversion systems, depending on the solar incidence and other cost factors at their location.[29,30] Figure 6.2 illustrates in a general fashion the cost at which photovoltaic power systems should become competitive with conventional generators as a function of the price of primary energy. Market penetration should begin when the price of photovoltaics decreases from the present large order price of $8–10/$W_p$ to about $2.5/$W_p$.

The first large power applications will probably occur where additional power is needed to supplement hydroelectric plants and where additional peaking capacity is needed in the sunbelt. Because hydroelectric power can be stored by simply not releasing the water through the turbines, photovoltaic power could be used in a hybrid system

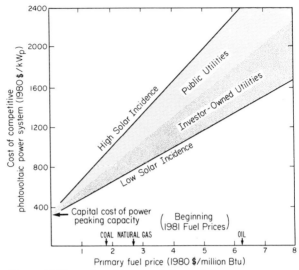

Figure 6.2 Illustration of the cost of photovoltaic power systems necessary to be competitive with thermoelectric central power systems as a function of primary fuel price. (Based on data in references 19, 25, 29, and 30.)

to provide electricity to the distribution grid during sunlit hours, thus conserving hydropower for use during nonsunlit hours. Hydroelectric generation accounts for about 14% of the electricity used in the United States. If photovoltaics were provided to supplant the hydroelectric generation during daylight hours so that it could be conserved for use in nonsunlit hours, about 85,000 MW_p would be required. Allowing for a gradual penetration of this market and replacement every 20 yr, an annual photovoltaic array volume of perhaps 500–4000 MW_p may be projected. Oil is used as the primary fuel for about 19% of the thermally generated electricity in the United States, or about 16% of the total. Because it is typically used for peaking capacity, it should be amenable to partial substitution by photovoltaics. Again, an annual market of 500–4000 MW may be anticipated.

Indeed, the entire 600,000-MW market may be considered as a potential market for photovoltaics as the price is reduced. However, because of the intermittency of sunshine and the magnitude of capital investment already in conventional facilities, a maximum penetration of 20% might be expected even after photovoltaics become cost competitive with conventional power generation in most locations. In 1980 terms, this would be a 120,000-MW potential. Allowing for a 5% penetration per year and a 20-yr replacement cycle, this would lead to a 6,000-MW market. Because the variations in demand require excess capacity and an average load factor of about 0.6 in conventional systems, a photovoltaic system operating at a load factor of about 0.2 (ratio of average insolation to peak insolation) will require about three times as much peak power as is required in the conventional system, so that an annual market of 18,000 MW_p may be calculated. Figure 6.3 shows a graphical representation of the approximate photovoltaic market size as a function of price that may be reasonably addressed by US manufacturers.

It is evident from figure 6.3 that lower prices for photovoltaic conversion systems will make them competitive in an ever broadening market. Because the experience in this business has been accumulated almost entirely using one type of technology, the large-crystal silicon solar cell, an analysis can be made of the change of price to expect in the future as a function of production experience in making similar photovoltaic arrays. Although market data are sketchy, an examination of references 4, 11, 12, 14, 16, 31, and 32 indicates that arrays for terrestrial use have been sold at the quantities and prices (in constant 1980 dollars) shown in figure 6.4. Significant sales began in about 1973. Prior to that, some production experience had been gained in providing arrays for the US space program, but because this

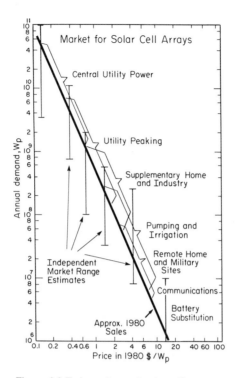

Figure 6.3 Estimated annual solar cell array market size as a function of price. Price is in 1980 dollars. Demand in W_p can be converted to square feet of arrays by the formula square feet = W_p ÷ efficiency ÷ 100 $W_p/$ ft². The price per peak watt can be converted to price per square foot by the formula price/ft² = price/W_p · Efficiency · 100 W_p/ft².

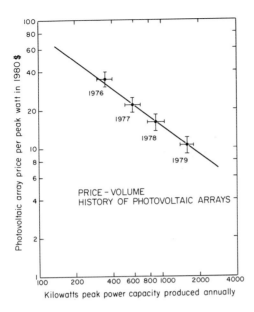

Figure 6.4 Historical sales of silicon solar cell arrays as a function of price.

demand was not indicative of a competitive terrestrial market, it is not included in the graph. However, the space program did enable the research, development, and manufacturing that led to the beginning of the terrestrial photovoltaic power business.

Over the 4-yr period from the end of 1975 to the end of 1979, the production of photovoltaic arrays increased about 65%/yr as the price decreased about 30%/yr in constant dollar value. By using this information, a classical learning or experience curve[33] can be constructed to predict future prices as production experience is accumulated. Because of the normal development of improved work flow, tooling, and manufacturing methods that come with experience, less waste is encountered and fewer labor hours per unit are needed. By extending figure 6.4 back to 1973, the cumulative production during the 1973–1975 period can be estimated at 400 kW$_p$, about the volume estimated in other studies.[4] An experience curve is obtained by plotting the cumulative average price per unit versus the cumulative units produced. This is shown in figure 6.5 using constant 1980 dollars for prices. The slope of the graph indicates that over the beginning years of the photovoltaic array business the cumulative average price decreased by 27% every time the cumulative production doubled. This is referred to as a 73% learning curve. As a point of interest, the learning curve in the aircraft industry is about 80%, and in the electronics industry it is about 70%. If the trends illustrated in figures 6.4 and 6.5 are extended, one finds that the price in 1980

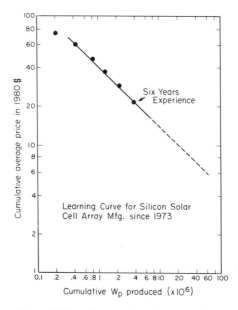

Figure 6.5 Experience or learning curve for manufacture of silicon solar cell arrays.

dollars falls to $1/W_p$ in 1988, when the annual volume is 145 MW_p. The cumulative production reaches 350 MW_p in that year. It is risky to extrapolate from so little data very far into the future. However, solar cell array manufacture is very labor intensive and hence subject to learning curve analysis. Furthermore, as seen in chapter 4, technological progress is driving the costs of the semiconductive materials in solar cells downward at a rate that indicates that structural costs in an array will probably be the limiting factor in cost reduction in the long term. This will probably happen when the cost reaches about $2/ft^2$, or 20¢/W_p, in a 10% efficient array.

Photovoltaic Business Forecast

By summarizing and combining the trends in the factors that determine the demand for electricity from photovoltaic conversion systems, a business projection can be made. Referring to the systems dynamics model represented in figure 6.1, it is seen that the factors that increase this demand are increasing price of electricity and pollution from conventional generation; decreasing reliability of conventional generation; increasing overall demand for electricity; decreasing price of photovoltaic electricity; and increasing reliability of the photovoltaic energy system. The trends in _all_ of these factors portend an increasing business in photovoltaic energy conversion. Furthermore, the governmental stimulation in photovoltaic system research and development (figure 5.4) and the steep learning curve with increased production experience (figure 6.5) are very positive driving forces.

The two most influencial factors are the price of electricity from photovoltaic systems and the size of the market that can be addressed at that price. At present, the price of arrays is about $10/W_p$ and that of complete systems varies from $10 to $25/W_p$. The annual business in total systems is $50–70 million. The projected future price of photovoltaic systems is illustrated in figure 5.9. The annual market as a function of price is estimated in figure 6.3. When the price projection as a function of time is convoluted with the annual demand as a function of price, a sales potential in 1980 dollars as a function of time is obtained as shown in figure 6.6. This projection also incorporates a 3.5% annual increase in demand for electricity. The lower part of the graph represents the potential business from solar cell arrays alone,

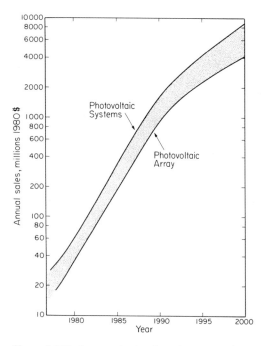

Figure 6.6 Business projection for solar array and system sales until the year 2000. Sales are all in 1980 dollars. A 3.5% annual growth in the market for electricity is incorporated.

and the higher portion represents the total business associated with an average system—some including storage, some only power conditioning, and others both storage and conditioning. By 1990 this should be a billion-dollar business with a high growth rate extending well into the twenty-first century.

Business success will depend first on the manufacturers' ability to drive costs downward so that larger market segments can be addressed. Second, it will depend on marketing and distribution channel efficiencies, particularly as they affect the developing countries, where favorable solar radiation conditions abound and relatively few conventional thermoelectric generation systems are in place. Third, it will depend on the installation service, maintenance, and reliability of photovoltaic system operation. And finally, business success will depend on the financing options offered relative to the competitive means for buying electricity. In the long term, competition for business may depend on product refinement that provides longer-lived arrays, more versatile modules, enhanced styling, greater ease of installation, and increased efficiency. Although a 100% increase in efficiency would take the simple flat plate silicon cell arrays almost to the theoretical limit, small increments in this aim would make significant differences in areas needed and potential uses for photovoltaics. For example, a doubling of efficiency would mean the average residential roof area may support electrical heating as well as the other electrical needs of the household. Lower space requirements, increased portability, and less visual degradation of the environment will accompany photovoltaic conversion efficiency gains.

Summary

The availability of terrestrial sunlight is abundantly sufficient to meet the energy needs of the United States and most parts of the world. The advantages of its long-term and widespread availability and pollution-free utilization are offset somewhat by its short-term fluctuations and the capital investment needed to provide for its collection and storage in a form acceptable for use during hours without sunlight.

The scientific principles upon which the photovoltaic effect is based allow that up to about 25% of collected sunlight can be converted directly into electricity at ambient temperatures with relatively simple solar cell designs. The state of the art permits about 12% efficiency with silicon devices having 20 yr life with negligible degradation. Today, 1981, the cost of silicon solar cell arrays is 10–20 times higher than that necessary to compete economically with thermoelectric generation by large central stations. It is 2–4 times too expensive to compete with oil-fired peaking capacity. The cost of electrical storage necessary to provide standalone photovoltaic energy systems is about 2–4 times too high. Power conditioning to provide alternating current at appropriate voltages is perhaps three times too expensive. Photovoltaic converters, which are most advantageous for remote hard-to-service locations (such as space) have gained acceptance (such as in cathodic protection, novelty uses, and communications) as more reliable and cost effective replacements of batteries and engine-driven generators. Growth in production has been about 65%/yr as the price has fallen 30%/yr over the last 5 yr. At present, it is a $50–70 million business.

The expected technological development of photovoltaic cell arrays, battery storage, and power-conditioning systems portends that photoelectricity will become cost competitive for irrigation, utility peak power, and general supplementary use by about 1988, and for general power production by 2000. Annual production of solar cell arrays should go from about 2.5 MW_p in 1980 (250,000 ft²) to about 4000 MW_p (400 million ft²) by the year 2000—a compounded growth rate of 45%/yr. Although the timetable is difficult to define precisely, demand for a stable supply of electricity at a reasonable price will stimulate the innovation required to make photovoltaic energy conversion a practical method for power generation. There is little doubt that the direct use of sunlight to provide electricity will become a multibillion dollar business in this century. Entrepreneurial opportunities will abound.

References

1. J. P. Holdren, G. Morris, and I. Mintzer, *Ann. Rev. Energy* 5 (1980): 252.

2. E. Robertson, *The Solarex Guide to Solar Electricity* (Washington, DC: Solarex Corp., 1979).

3. H. Kelly, *Energy II: Use, Conservation, and Supply*, P. H. Abelson and A. L. Hammond, eds. (Washington, DC: Am. Assoc. Adv. Sci., 1978), p. 151.

4. D. Costello and P. Rappaport, *Ann. Rev. Energy* 5 (1980): 335.

5. *Characterization of the Present Worldwide Photovoltaic Power Systems Market* (McLean, VA: BDM Corp., May 1977), p. 1.

6. *Photovoltaic Power Systems Market Identification and Analysis* (Warrenton, VA: Intertechnology Corp., August 1977).

7. *Department of Defense Photovoltaic Energy Conversion*

Systems Market Inventory and Analysis Summary Volume (McLean, VA: BDM Corp., 1977), p. 17.

8. *Final Report: Conceptual Design and Systems Analysis of Photovoltaic Power Systems*, Vol. I, *Executive Summary* (Pittsburgh: Westinghouse Corp., April 1977), p. 45.

9. P. D. Maycock and G. F. Wakefield, *Proc. 11th IEEE Photo. Spec. Conf.* (Scottsdale, AZ: IEEE, 1975).

10. R. Scott, *The Solar Market, Proc. Symp. on Competition in the Solar Energy Industry* (Washington, DC: Federal Trade Commission, June 1978), p. 295.

11. O. Merrill, *Photovoltaic Power Systems Market Identification* (McLean, VA: BDM Corp., 1977).

12. H. Liers, *Photovoltaic Energy Technology Market Analysis* (Warrenton, VA: Intertechnology Solar Corp., 1979).

13. Intertechnology Solar Corp., Arthur D. Little Inc. Report, P. E. Glaser, *Progress in Photovoltaics* (Cambridge, MA: Arthur D. Little Inc., May 1980), p. 3.

14. W. D. Metz and A. L. Hammond, *Solar Energy in America* (Washington, DC: Am. Assoc. Adv. Sci., 1978), p. 57.

15. *Solar Photovoltaic Energy Research, Development, and Demonstration Act of 1978*, Public Law 95-590, Section 11a.

16. Science Applications, Inc., *Characterization and Assessment of Potential European and Japanese Competition in Photovoltaics* (Golden, CO: Solar Energy Research Institute, DOE Subcontract No. XP-9-3251-1, 1979), p. 3–15.

17. Spectrolab, Inc., *Photovoltaic Systems Concept Study* (Washington, DC: ERDA, 1977).

18. Westinghouse Electric Corp., *Conceptual Design and Analysis of Photovoltaic Systems* (Washington, DC: ERDA, 1977).

19. General Electric Corp., *Requirements Assessment of Photovoltaic Electric Power System* (Palo Alto, CA: Electric Power Research Institute, EPRI ER-685-SY Vol. 1, Summary Report, 1978).

20. General Electric Corp., *Regional Conceptual Design and Analysis of Photovoltaic Systems* (Albuquerque, NM: Sandia Laboratories DOE, 1979).

21. Aerospace Corp., *Mission Analysis of Photovoltaic Solar Energy Conversion: Major Missions for 1986–2000* (Washington, DC: DOE, 1977).

22. D. Costello and D. Posner, *Solar Cells* 1 (1979): 37.

23. E. DeMeo and P. Bos, *Perspectives on Utility Central Station Photovoltaic Applications* (Palo Alto, CA: Electric Power Research Institute, EPRI ER-589-SR Special Report, 1978).

24. A. J. Cox, *The Economics of Photovoltaics in the Commercial, Institutional, and Industrial Sectors* (Cambridge, MA: MIT, 1980).

25. J. O. Bradley and D. R. Costello, *Solar Energy* 19 (1977):701.

26. V. Evtuhov, *Solar Energy* 22 (1979): 427.

27. E. A. DeMeo and P. B. Bos, *Solar Energy* 21 (1978): 177.

28. E. L. Ralph, *Solar Energy* 13 (1972): 326.

29. G. J. Jones, *Proc. DOE Ann. Photo. Program Rev. for Tech. and Market Dev.* (Hyannis, MA: DOE, 1980), p. 109.

30. S. L. Leonard, *Proc. DOE Ann. Photo. Program Rev. for Tech. and Market Dev.* (Hyannis, MA: DOE, 1980), p. 465.

31. BDM Corp., *Photovoltaic Incentive Options, Preliminary Report* (McLean, VA: BDM Corp. W-78-184-TR, 1978).

32. Booz, Allen and Hamilton, Inc., *Assessment of Solar Photovoltaic Industry Markets and Technologies, Draft Report* (Bethesda, MD: Booz, Allen and Hamilton, Inc., 1978).

33. R. N. Anthony, *Management Accounting* (Homewood, IL: Richard D. Irvin, Inc., 1970), 4th ed., pp. 467–469.

Appendix: Glossary of Terms and Conversions

am (air mass) number A term describing the spectrum and intensity of sunlight. AM0 indicates that the sunlight has not passed through the atmosphere, about 136 mW/cm^2. AM1 indicates that the sunlight has passed through one vertical atmospheric thickness with the attendant light absorption and scattering, about 100 mW/cm^2. AM2 refers to sunlight passed through two vertical thicknesses of the atmosphere; that is, it passed through the atmosphere at an angle.

British thermal unit (Btu) The energy required to raise the temperature of 1 lb of water 1° Fahrenheit: 2.93×10^{-4} kWh; 1054.8 J (joules).

cell efficiency The ratio of output electrical energy to the input solar energy falling on a solar cell times 100. This is usually determined by shining a simulated AM1 light source of 100-mW/cm^2 intensity over the entire surface of the cell and relating the wattage output to the total input.

full sun The power density received at the surface of the earth at noon on a clear day. About 100 mW/cm^2. Lower light intensities may be described as a fraction of a sun. Higher levels encountered in concentration solar cells may be described as multiple suns.

horsepower The power of 745.6 W or 0.70696 Btu/sec.

insolation The intensity of sunlight reaching a given area, usually expressed in milliwatts per square centimeter. This may express average insolation in referring to solar energy falling on different regions of the country.

kilowatt-hour 1000 W of power used for a 1-h period: 3413 Btu; 3.6×10^6 J.

photon A discrete quantum of energy corresponding to a given wavelength of light. The relation to wavelength can be calculated using the equation photon energy (eV) = 1.234 (μm eV) $\div$ wavelength (μm).

photovoltaic cell Solar cell, that is, a device that converts light into electrical energy by direct absorption followed by separation of positive and negative photocarriers.

power density The power collectable over a given area as used in solar energy. The power exertable by a given weight of device as in a battery. Terms are typically mW/cm^2 or kW/lb.

solar constant The rate at which solar energy is received from the sun just outside the earth's atmosphere at an average distance of the earth from the sun: 136 mW/cm^2.

watt A power of 1 J, or 10^7 erg/sec.: 9.4827×10^{-4} Btu/sec; 1.341×10^{-3} hp.

Conversions

Units of Weight
short ton = 907.185 kg = 2000 lb (avoir.)
long ton = 1016.047 kg = 2240 lb (avoir.)
metric ton = 1000 kg = 2205 lb (avoir.)
pound (avoir.) = 453.59 g

Units of Length
mile = 5280 ft = 1.60934 km
meter (m) = 1.0936 yd = 39.37 in.
yard = 36 in. = 91.44 cm
foot (ft) = 12 in. = 30.48 cm
inch (in.) = 2.54 cm

micrometer (μm) = 10^{-6} m = 1000 nm

Units of Area
ft² = 144 in.² = 929.03 cm² = 0.0929 m²
m² = 10.764 ft²
yd² = 9 ft² = 0.836 m²
acre = 4046.87 m² = 43,560 ft²
mile² = 27,878,400 ft² = 640 acres = 2.6 km²
km² = 0.3861 miles² = 1.0764×10^7 ft²
United States = 3.6×10^6 miles² = 2.3×10^9 acres

Units of Time
year (yr) = 365.25 days = 8766 h = 5.2596×10^5 min
         = 3.15576×10^7 sec
day = 24 h
hour (h) = 60 min
minute (min) = 60 sec

Units of Energy
Btu = 2.93×10^{-4} kWh = 1054.8 J
joule (J) = 2.778×10^{-7} kWh
erg = 10^{-7} J
kilowatt-hour (kWh) = 3413 Btu = 3.6×10^6 J

Units of Power
horsepower (hp) = 745.6 W = 0.70696 Btu/sec
kilowatt (kW) = 1.341 hp = 1000 W
watt = 9.4827×10^{-4} Btu/sec = 1 J/sec = 10^7 erg/sec

Units of Primary Fuels
crude oil (barrel) = 42 gallons = 5,800,000 Btu (ave.)
coal (short ton) = 26,000,000 Btu (ave.)
natural gas (cubic foot) = 1070 Btu
electricity (kilowatt-hour) = 3412 Btu

Bibliography

I. Abrahamsohn, US Patent 3,376,163 (2 April 1968).

Aerospace Corp., *Mission Analysis of Photovoltaic Solar Energy Conversion: Major Missions for 1986–2000* (Washington, DC: DOE, 1977).

C. W. Allen, *Quart. J. Roy. Met. Soc.* 84 (1958): 307

S. W. Angrist, *Direct Energy Conversion* (Boston: Allyn and Bacon, 1965), p. 326.

R. N. Anthony, *Management Accounting* (Homewood, IL: Richard D. Irvin, Inc., 1970), 4th ed., pp. 467–469.

R. A. Arndt, J. F. Allison, J. G. Haynos, and A. Meulenberg, *Proc. 11th IEEE Photo. Spec. Conf.* (Scottsdale, AZ: IEEE, 1975), p. 40.

Associated Universities, Inc., *Reference Energy Systems and Resource Data for Use in the Assessment of Energy Technologies* (Report to US Office of Science and Technology, under Contract OST-30; Document AET-8, April 1972).

D. H. Auston, C. M. Surko, T. N. C. Venkatesan, R. E. Slusher, and J. A. Golovchenko, *Appl. Phys. Lett.* 33 (1978): 130.

P. Baeri, S. U. Campisano, and E. Rimini, *Appl. Phys. Lett.* 33 (1978): 137.

B. Baker, *Fuel Cell Systems—II, Advances in Chemistry Series*, R. F. Gould, ed. (Washington, DC: American Chemical Society

Publications, 1969).

A. Barna, P. B. Barna. G. Rodnoczi, L. Toth, and P. Thomas, *Phys. Stat. Sol.* 41 (1977): 81.

A. Barnett, J. A. Brogagnolo, R. B. Hall, J. E. Phillips, J. D. Meakin, *Proc. 13th IEEE Photo. Spec. Conf.* (New York: IEEE, 1978), p. 419.

G. O. Barney, *The Global 2000 Report to the President* (Washington, DC: US Government Printing Office, vol. 1, 1980).

G. Bassak, *Electronics* (14 August 1980): 43.

BDM Corp., *Photovoltaic Incentive Options, Preliminary Report* (McLean, VA: BDM Corp. W-78-184-TR, 1978).

R. L. Bell, *Solar Energy* 23 (1979): 203.

D. Bernatowicz and H. Brandhorst, Jr., "The Degradation of Cu_2S-CdS Thin Film Solar Cells Under Simulated Orbital Conditions." *Proc. 8th IEEE Phot. Spec. Conf.* (Seattle, WA: IEEE, 1970), p. 24.

P. J. Bernstein, Newhous News Service, *Times Union* (Rochester, NY: 31 December 1980): 1A.

K. W. Böer, *Chem. Eng. News* (29 January 1973): 12.

K. W. Böer, *Proc. 11th IEEE Photo. Spec. Conf.* (Scottsdale, AZ: IEEE, 1975), p. 514.

K. W. Böer, *Proc. Joint Conf. Am. Sect. ISES and SES Can. Inc. Winnipeg* 1 (August 1976): 264.

E. C. Boes and B. D. Shafer, *Proc. DOE Ann. Photo. Program Rev. for Tech. and Market Dev.* (Hyannis, MA: DOE, 1980), pp. 29, 46.

R. Boll and R. Bhada, *Energy Conversion* 8 (1968): 3.

Booz, Allen and Hamilton, Inc., *Assessment of Solar Photovoltaic Industry Markets and Technologies, Draft Report* (Bethesda, MD: Booz, Allen and Hamilton, Inc., 1978).

J. O. Bradley and D. R. Costello, *Solar Energy* 19 (1977):701.

J. B. Brinton, *Electronics* (11 September 1980): 39.

M. H. Brodsky, M. A. Frisch, J. F. Ziegler, and W. A. Landford, *Appl. Phys. Lett.* 30 (1977): 561.

M. L. Brown, *Proc. DOE Ann. Photo. Program Rev. for Tech. and Market Dev.* (Hyannis, MA: DOE, 1980), p. 477.

W. C. Brown, *Microwave Power Transmission in the Satellite Solar Power Station System* (Raytheon Company Technical Report ER72–4038, 1972).

A. Bruckner, II, W. Fabrycky, and J. Shamblin, *IEEE Spectrum* (April 1968): 101.

E. Bucher, *Appl. Phys.* 17 (1978):1.

Bureau of the Census, *Projections of the Population of the United States 1975 to 2000* (Washington, DC: US Department of Commerce, 1975).

D. E. Burk, J. B. Dubow, and J. R. Sites, *Device Research Conf.* (Salt Lake City, UT: 1976).

Business Bulletin. *Wall Street Journal* (New York, 4 September 1980): 1.

R. S. Caputo, "Toward a Solar Civilization," *Solar Power Plants: Dark Horse in the Energy Stable,* R. H. Williams, ed., (Cambridge, MA: MIT Press, 1978), pp. 73–93.

H. C. Card and E. S. Yang, *Appl. Phys. Lett.* 29 (1976): 51.

D. E. Carlson, *IEEE Trans. Electron Dev.* ED-24 (1977): 449.

D. E. Carlson, US Patent No. 4,064,521 (1977).

D. E. Carlson, "Amorphous Silicon," to be published in *Progress in Crystal Growth and Characterization,* 1981.

D. E. Carlson and C. W. Magee, *Appl. Phys. Lett.* 33 (1978):81.

D. E. Carlson and C. W. Magee, *2nd E. C. Photovoltaic Solar Energy Conf. Berlin* (Dardrecht, Holland: Reidel, 1979).

D. E. Carlson, R. W. Smith, G. A. Swartz, and A. R. Triano, "5.5% *p-i-n* Amorphous Silicon Solar Cells," *Extended Abstracts*, vol. 80–2 (Hollywood, FL: Electrochemical Society Mtg., October 1980), p. 1428.

D. E. Carlson, C. R. Wronski, J. I. Pankove, P. J. Zanzucchi, and D. L. Staebler, *RCA Review* 39 (1977): 211.

A. N. Casperd and R. Hill, *Solar Cells* 1 (1980): 347.

J. A. Castle, *Proc. 12th IEEE Photo. Spec. Conf.* (Baton Rouge, LA: IEEE, 1976), p. 751.

K. C. Chang, A. Heller, B. Schwartz, S. Menezes, and B. Miller, *Science* 196 (1977): 1097.

D. M. Chapin, C. S. Fuller, and G. L. Pearson, *J. Appl. Phys.* 25 (1954): 676.

H. K. Charles, Jr., and A. P. Ariotedjo, *Solar Energy* 24 (1980): 329.

G. Cheek and R. Mertens, *Solar Cells* 1 (1980): 405.

G. Cheek, N. Inove, S. Goodnick, A. Genis, C. Wilmsen, and J. DuBow, *Appl. Phys. Lett.* 33 (1978): 643.

W. R. Cherry, *Proceedings of the 13th Annual Power Sources Conference* (Atlantic City, NJ: IEEE, May 1959).

R. C. Chittick, J. H. Alexander, and H. F. Sterling, *J. Electrochem. Soc.* 116 (1969): 77.

S. S. Chu, T. L. Chu, Y. T. Lee, *Proc. 14th IEEE Photo. Spec. Conf.* (San Diego: IEEE, 1980).

T. F. Ciszek, *Mat. Res. Bull.* 7 (1972): 731.

G. A. N. Connell and J. R. Pawnk, *Phys. Rev. B* 13 (1976): 787.

I. E. Cook, *Sci. Amer.* 225 (1971): 137.

D. Costello and D. Posner, *Solar Cells* 1 (1979): 37.

D. Costello and P. Rappaport, "The Technological and Economic

Development of Photovoltaics," *Ann. Rev. Energy*, J. M. Hollander, M. K. Simmons, and D. O. Wood, eds. (Palo Alto, CA: Annual Reviews Inc., 1980), pp. 335–356.

A. J. Cox, *The Economics of Photovoltaics in the Commercial, Institutional, and Industrial Sectors* (Cambridge, MA: MIT, 1980).

C. H. Cox, *Proc. DOE Ann. Photo. Program Rev. for Tech. and Market Dev.* (Hyannis, MA: DOE, 1980), p. 281.

R. Crabb, "Status Report on Thin Silicon Solar Cells for Flexible Arrays" *Solar Cells* (New York: Gordon and Breach, 1971), pp. 35–50.

C. Currin, K. Ling, E. Ralph, W. Smith, and R. Stirn, "Feasibility of Low Cost Silicon Solar Cells," *Proc. 9th IEEE Photo. Spec. Conf.* (Maryland: IEEE, May 1972).

F. Daniels, *Solar Energy* 6 (1962): 78.

J. David, S. Martinuzzi, F. Cabane-Brouty, J. Sorbier, J. Mathieu, J. Roman, and J. Bretzner, "Structure of CdS-Cu$_2$S Heterojunction Layers," *Solar Cells* (New York: Gordon and Breach, 1971), pp. 81–94.

B. T. Debney and J. R. Knight, *Contemp. Phys.* 19 (1978): 25.

E. A. DeMeo, Electric Power Research Institute Report No. ER-188 (February 1976).

E. A. DeMeo, and P. Bos, *Perspectives on Utility Central Station Photovoltaic Applications* (Palo Alto, CA: Electric Power Research Institue, EPRI ER-589-SR Special Report, 1978).

E. A. DeMeo and P. B. Bos, *Solar Energy* 21 (1978): 177.

F. DeMichelis and E. Minetti-Mezzetti, *Solar Cells* 1 (1980): 395.

W. Dickter, *Proc. DOE Ann. Photo. Program Rev. for Tech. and Market Dev.* (Hyannis, MA: DOE, 1980), p. 407.

T. L. Dinwooie, *Flywheel Storage for Photovoltaics: An Eco-*

nomic Evaluation of Two Applications (Cambridge, MA: MIT Energy Laboratory, February 1980).

R. C. Dorf, *Energy, Resources, and Policy* (Reading, MA: Addison-Wesley, 1978).

D. L. Douglas and J. R. Birk, *Ann. Rev. Energy* 5 (1980): 61.

T. D. Duchesneau, Federal Trade Commission Economic Report, *Interfuel Substitutability in the Electric Sector of the U.S. Economy* (Washington, DC: US Government Printing Office, 1972).

Edison Electric Institute, *Statistical Year Book of the Electric Utility Industry* (New York, 1978).

Edison Electric Institute, *Source and Disposition of Electricity* (New York: Edison Electric Institute, Volume 48, No. 6, 18 September 1980).

H. Ehrenreich, *Principal Conclusion of the American Physical Society's Study Group on Solar Photovoltaic Energy Conversion* (New York: Am. Phys. Soc., 1979).

H. Ehrenreich and J. H. Martin, *Phys. Today* 32 (September 1979): 25.

J. A. Eibling, *Solar Energy: An Assessment for Business*, B-TIP Rev. No. 2 (Columbus, OH: Battelle Memorial Institute, 1979), pp. 4, 18.

J. F. Elliott, "Photovoltaic Energy Conversion," *Direct Energy Conservation*, G. W. Sutton, ed. (New York: McGraw-Hill, 1966), pp. 1–37.

Energy Group, *Capital Resources for Energy through the Year 1990* (New York: Bankers Trust Company, 1976).

ERDA-49 *National Solar Energy Research, Development and Demonstration Program* (Washington, DC: US Government Printing Office, Superintendent of Documents, June 1975).

D. L. Evans and L. W. Florschuetz, *Solar Energy* 20 (1978): 37.

F. C. Eversteyn, *Phillips Res. Rept.* 29 (1976): 45.

V. Evtuhov, *Solar Energy* 22 (1979): 427.

Exxon Company, U.S.A., *Energy Outlook 1976–1990* (December 1975).

Federal Energy Administration, *National Energy Outlook* (Washington, DC: US Government Printing Office, 1976).

T. Feng, A. K. Ghosh, and C. Fishman, *Appl. Phys.Lett.* 35 (1979): 266.

T. Feng, A. K. Ghosh, and C. Fishman, *J. Appl. Phys.* 50 (1979): 4972.

D. L. Feucht, "Photovoltaic R and D. Program Overview," *Proceedings of the DOE Annual Photovoltaics Program Review for Technology and Market Development* (Massachusetts, 1980), Conf-8004101, NTIS, US Dept. of Commerce, Springfield, VA.

A. Flat and A. G. Milnes, *Solar Energy* 23 (1979): 289.

A. Flat and A. G. Milnes, *Solar Energy* 25 (1980): 283.

J. W. Forrester, *Industrial Dynamics* (Cambridge, MA: MIT Press, 1961).

J. W. Forrester, *Principles of Systems* (Cambridge, MA: Wright-Allen Press, 2nd preliminary ed., 1972).

J. G.Fossum and E. L. Burgess, *Proc. 12th IEEE Photo. Spec. Conf.* (Baton Rouge, LA: IEEE, 1976), p. 737.

J. G. Fossum and F. A. Lindholm, *IEEE Trans. Elec. Dev.* ED-24 (1977): 325.

A. Fujishima and K. Honda, *Nature* 238 (1972): 38

H. Garfinkel and R. N. Hall, *Floating Substrate Process*, ERDA/JPL 954350–76/1, *First Q. Prog. Rept.* (1976).

General Electric Corp., *Requirements Assessment of Photovoltaic Electric Power System* (Palo Alto, CA: Electric Power Research Institute, EPRI ER-685-SY Vol. 1, Summary Report. 1978).

General Electric Corp., *Regional Conceptual Design and Analysis of Photovoltaic Systems* (Albuquerque, NM: Sandia Laboratories DOE, 1979).

General Electric Space Division, *Applied Research on Energy Storage and Conversion for Photovoltaic and Wing Energy Systems*, Final Report, January 1978.

A. K. Ghosh and T. Feng, *J. Appl. Phys.* 44 (1973): 2781.

A. K. Ghosh, T. Feng, and H. P. Maruska, *Solar Cells* 1 (1980): 421.

A. K. Ghosh, C. Fishman, and T. Feng, *J. Appl. Phys.* 51 (1980): 446.

A. K. Ghosh, D. Morel, T. Feng, R. Shaw, and C. Rowe, *J. Appl. Phys.* 45 (1974): 230.

P. E. Glaser, *Space Resources to Benefit the Earth, Third Conference on Planetology and Space Mission Planning* (The New York Academy of Sciences, October 1970).

W. Glasgall, Associated Press, *Times Union* (Rochester, NY: 10 September 1980): 6D.

R. B. Godfrey and M. A. Green, *Appl. Phys. Lett.* 33 (1978): 637.

R. B. Godfrey and M. A. Green, *Appl. Phys. Lett.* 34 (1979): 790.

A. Goetzberger and W. Breubel, *Appl. Phys.* 14 (1977): 123.

R. Gold, "Current Status of GaAs Solar Cells," Transcript of Photovoltaic Specialists Conference, Vol. 1, *Photovoltaic Materials, Devices and Radiation Damage Effects* (DDC No. AD412819, July 1963).

R. M. Goody, *Atmospheric Radiation* (Oxford: Clarendon Press, 1964), pp. 417–426.

C. Goradia, R. Ziegman, and B. L. Sater, *Proc. 12th IEEE Photo. Spec. Conf.* (Baton Rouge, LA: IEEE, 1976), p. 781.

P. Gotlieb, "Alternative Fuels," *Engineering Bulletin* 51 (1980): 34.

J. F. Graczyk, *Phys. Stat. Sol.* 55 (1979): 231.

C. D. Grahm, S. Kulkarni, G. T. Noel, D. P. Pope, B. Pratt, and M. Wolf, *Hot Forming of Silicon*, ERDA/JPL 954506-76/1, *Q. Report No. 1* (1976).

J. Griffin, *Bell J. Econ. Mgmt.* 5 (1974): 515.

J. A. Grimshaw and W. G. Townsend, *Solar Cells* 2 (1980): 55.

Gulf Oil Corporation, *Annual Report* (1979), pp. 4,5.

Gulf and Western, *Wall Street Journal* (6 June 1980): 11.

G. M. Haas and S. Bloom, *Proc 11th IEEE Photo. Spec. Conf.* (Scottsdale, AZ: IEEE, 1975), p. 256.

D. S. Halacy, *The Coming Age of Solar Energy* (New York: Avon Books, 1975), pp. 23–38.

D. E. Hall, J. A. Eckert, N. N. Lichtin, and P. D. Wildes, *J. Electrochem. Soc.* 123 (1976): 1705.

R. Halsted, *J. Appl. Phys.* 28 (1957): 1131.

J. J. Hanak, *Solar Energy* 23 (1979): 145.

J. D. Heaps, R. B. Maciolek, W. B. Harrison, and H. A. Wolner, *Dip-Coating Process*, ERDA/JPL 954356, *Q. Report No. 1* (1975).

A. Heller, K C. Chang, and B. Miller, *Semiconductor Liquid-Junction Solar Cells, Proceedings*, vol. 77-3 (Princeton: Electrochemical Society, 1977), pp. 54–66.

S. T. Henderson, *Daylight and Its Spectrum* (New York: American Elsevier, 1970).

J. P. Holdren, G. Morris, and I. Mintzer, *Ann. Rev. Energy* 5 (1980): 252.

B. Horovitz, *Industry Week* (26 May 1980): 68.

B. Horovitz, *Industry Week* (26 May 1980): 70.

B. Horovitz, *Industry Week* (19 September 1980): 80.

R. W. Hosken, *Electro. Optical Sys. Design* (January 1975): 32.

H. J. Hovel, "Solar Cells," *Semiconductors and Semimetals*, vol. II, A. C. Beer and R. K. Willardson, eds. (New York: Academic Press, 1975).

H. J. Hovel, *Solar Energy* 19 (1977): 605

M. Igbal, *Solar Energy* 22 (1979): 81.

M. Igbal, *Solar Energy* 23 (1979): 169.

Intertechnology Solar Corp., Arthur D. Little Inc. Report, P. E. Glaser, *Progress in Photovoltaics* (Cambridge, MA: Arthur D. Little Inc., May 1980). p.3

L. W. James and R. L. Moon, *Appl. Phys. Lett.* 26 (1975): 467.

L. W. James and R. L. Moon, *Proc. 11th IEEE Photo. Spec. Conf.* (Scottsdale, AZ: IEEE, 1975), p. 402.

P. O. Jarvinen, *Proc. DOE Ann. Photo. Program Rev. for Tech. and Market Dev.* (Hyannis, MA: DOE, 1980), p. 205.

J. Javetski, *Electronics* 52 (July 1979): 105.

Joint Committee on Atomic Energy, *Certain Background Information for Consideration When Evaluating the National Energy Dilemma* (Washington, DC: US Printing Office, 1973).

G. J. Jones, *Proc. DOE Ann. Photo. Program Rev. for Tech. and Market Dev.* (Hyannis, MA: DOE, 1980), p. 109.

J. F. Jordan, *Proc. 11th IEEE Photo. Spec. Conf.* Scottsdale, AZ: IEEE, 1975), p. 508.

S. Kar, K. Rajeshwar, P. Singh, and J. DuBow, *Solar Energy* 23 (1979): 129.

H. Kelly, *Energy II: Use, Conservation, and Supply*, P. H.

Abelson and A. L. Hammond, eds. (Washington, DC: Am. Assoc. Adv. Sci., 1978), p. 151.

H. Kelly, "Photovoltaic Power Systems: A Tour through the Alternatives," *Energy II: Use, Conservation, and Supply*, P. H. Abelson and A. L. Hammond, eds. (Washington, DC: Am. Assoc. Adv. Sci., 1978), pp. 151–160.

H. Kelly, *Science* 199 (1978): 634.

T. M. Klucher, *Solar Energy* 23 (1979): 111.

J. H. Krenz, *Energy: From Opulence to Sufficiency* (New York: Praeger, 1980), pp. 12, 13.

H. Landsberg, H. Lippmann, Kh. Paffen, and C. Troll, *World Maps of Climatology* (New York: Springer-Verlag, 2nd ed., 1965).

S. L. Leonard, *Proc. DOE Ann. Photo. Program Rev. for Tech. and Market Dev.* (Hyannis, MA: DOE, 1980), p. 465.

A. J. Lewis, G. A. N. Connell, W. Paul, J. R. Pawlik, and R. J. Temkin, *Proc. Int. Conf. on Tetrahedrally Bonded Amorphous Semiconductors* (New York: American Institute of Physics, 1974), p. 27.

N. N. Lichtin, *Photogalvanic Processes, Solar Power and Fuels* (New York: Academic Press, 1977), pp. 119–142.

N. N. Lichtin, *Chem. Tech.* (April 1980): 252.

H. Liers, *Photovoltaic Energy Technology Market Analysis* (Warrenton, VA: Intertechnology Solar Corp., 1979).

D. R. Lillington and W. G. Townsend, *Appl. Phys. Lett.* 28 (1976): 97.

J. Lindmayer, "Characteristics of Semi-Crystalline Silicon Solar Cells," *Proc. 13th IEEE Photo. Spec. Conf.* (New York: IEEE, 1978), pp. 1096–1100.

J. Loferski, *J. Appl. Phys.* 27 (1956): 777.

J. Loferski, *Proc. 12th IEEE Photo. Spec. Conf.* (New York: IEEE, 1976), p. 957.

D. J. Lootens, "The Nuclear Option," *Engineering Bulletin* 51 (1980): 18.

L. Lowe, *Electronics* (6 November 1980): 40.

A. Luque and E. Lorenzo, *Solar Energy* 22 (1979): 187.

A. Luque, E. Lorenzo, and J. M. Ruiz, *Solar Energy* 25 (1980): 171.

J. Lyman, *Electronics* (11 September 1980): 40.

L. Lyons and O. Newman, *Australian J. Chem.* 24 (1973): 13.

H. Macomber, *Proceedings of the ERDA Semiannual Solar Photovoltaic Program Review Mtg.* (Washington, DC: Energy Research and Development Administration, 1977), p. 68.

A. Madan, S. R. Ovshinsky, and E. Benn, *Philos. Mag.* 40 (1979): 259.

M. A. Maidique and B. Woo, *Tech. Rev.* (May 1980): 25.

J. Manassen, G. Hodes, and D. Cahen, *Semiconductor Liquid-Junction Solar Cells, Proceedings,* vol. 77–3 (Princeton: Electrochemical Society, 1977), pp. 34–37.

C. E. Mann and W. D. Bullock, "The Outlook for Coal," *Engineering Bulletin* 51 (1980): 11.

W. D. Marsh, *Requirements Assessment of Photovoltaic Power Plants in Electric Utility Systems* (Palo Alto, CA: Electric Power Research Institute, June 1978) EPRI ER-685-54, Vol. 1, Project 651–1, Summary Report.

M. Marshall, *Electronics* (3 January 1980): 102.

H. P. Maruska and A. K. Ghosh, *Solar Energy* 20 (1978): 443.

T. Matsushita and T. Maimine, *Proc. IEEE Int. Electron Dev. Mtg.* (Washington, DC: IEEE, 1975), p. 353.

P. D. Maycock, *Proc. DOE Ann. Photo. Program Rev. for*

Tech. and Market Dev. (Hyannis, MA: DOE, 1980), p. 2.

P. D. Maycock and G. F. Wakefield, *Proc. 11th IEEE Photo. Spec. Conf.* (Scottsdale, AZ: IEEE, 1975).

N. McBride, *Mid American Outlook* (Cleveland, OH: CleveTrust Corporation, 1980), pp. 10–12.

A. McDougall, *Fuel Cells* (New York: John Wiley and Sons, 1976).

T. C. McGill and C. A. Mead, *J. Vac. Sci. Technol.* 11 (1974): 122.

J. D. Meakin, B. Baron, K. W. Böer, L. Burton, W. Devaney, H. Hodley, J. Philips, A. Rothwarf, G. Storti, and W. Tseng, *6th Int. Solar Energy Soc. Conf.* (Winnipeg, 1976).

H. Meier, "Application of the Semiconductor Properties of Dyes: Possibilities and Problems," *Topics in Current Chemistry, Physical and Chemical Applications of Dyestuffs* (New York: Springer-Verlag, 1976) pp. 85–131.

J. A. Merrigan, *Sunlight to Electricity: Prospects for Solar Energy Conversion by Photovoltaics* (Cambridge, MA: MIT Press, 1975), pp. 1–13, 135.

O. Merrill, *Photovoltaic Power Systems Market Identification* (McLean, VA: BDM Corp., 1977).

V. Y. Merritt and H. J. Hovel, *Appl. Phys. Lett.* 29 (1976): 414.

W. D. Metz and A. L. Hammond, *Solar Energy in America* (Washington, DC: Am. Assoc. Adv. Sci., 1978), pp. 57, 103–123.

MIT Energy Laboratory and MIT Lincoln Laboratory, *Proposal for Solar-Powered Total Energy Systems for Army Bases* (Massachusetts Institute of Technology, July 1973).

R. M. Moore, *Solar Energy* 18 (1976): 225.

W. E. Morrow, Jr., *Tech. Rev.* 76 (1973): 31.

D. M. Mosher, R. E. Boese, and R. J. Soukup, *Solar Energy* 19 (1977): 91.

J. M. Mountz and H. Ti Tien, *Solar Energy* 21 (1978): 291.

N. Nakayama, H. Matsumoto, K. Yamaguchi, S. Ikegami, and Y. Hioki, *Jap. J. Appl. Phys.* 15 (1976): 2281.

National Petroleum Council, *U.S. Energy Outlook, An Initial Appraisal 1971–1985* (July 1971).

National Petroleum Council, *U.S. Energy Outlook, A Summary Report of the National Petroleum Council* (December 1972), p. 15.

National Petroleum Council, *U.S. Energy Outlook, A Report of the National Petroleum Council's Committee on U.S. Energy Outlook* (December 1972).

NSF/NASA Solar Energy Panel, *An Assessment of Solar Energy as a National Energy Resource* (University of Maryland, December 1972).

L. Olson, *Polysilicon Review Mtg.* (Golden, CO: Solar Energy Research Institute, June 1979).

S. R. Ovshinsky and A. Madan, *Nature* 276 (1978): 482.

M. A. Paesler, D. A. Anderson, E. C. Freeman, G. Moddel, and W. Paul, *Phys. Rev. Lett.* 41 (1978): 1492.

S. Yu. Pavelets and G. A. Fedorus, *Geliotek nika* 7 (1973): 3.

S. Yu. Pavelets and G. A. Fedorus, *Applied Solar Energy* 7 (1973): 1.

S. S. Penner and L. Icerman, *Energy: Demands, Resources, Impact, Technology, and Policy* (Reading, MA: Addison-Wesley, vol. 1, 1974), pp. 144–146.

E. A. Perez-Albuerne and Yuan-Sheng Tyan, *Science* 208 (1980): 902.

F. Pfisterer, H. W. Schock, and W. H. Bloss, *Proc. 15th IEEE Photo. Spec. Conf.* (Baton Rouge, LA: IEEE, 1976), p. 502.

J. G. Posa, *Electronics* (11 October 1979): 43.

J. G. Posa, *Electronics* (6 November 1980): 39.

D. L. Pulfrey, *Photovoltaic Power Generation* (New York: Van Nostrand Reinhold, 1978).

D. L. Pulfrey and R. F. McQuat, *Appl. Phys. Lett.* 24 (1974): 167.

K. Rajkanan, W. A. Anderson, and G. Rajeswaren, *IEEE Electron Device Mtg.* (Washington, DC: IEEE, December 1979).

E. L. Ralph, *Solar Energy* 13 (1972): 326.

P. Rappaport, *RCA Rev.* 20 (1959): 373.

P. Rappaport and J. Wysocki, *Acta Electronica* 5 (1961): 364.

P. Rappaport and J. Wysocki, "The Photovoltaic Effect," *Photoelectronic Materials and Devices*, S. Larach, ed. (New York: Van Nostrand, 1965), pp. 239–275.

R. D. News, *Industrial Research and Development* (January 1981): 43.

D. C. Reynolds, G. Leies, L. L. Antes, and R. E. Masbruger, *Phys. Rev.* 96 (1954): 533.

R. Riel, "Large Area Solar Cells Prepared on Silicon Sheet," *Proceedings of the 17th Annual Power Sources Conference* (Atlantic City, NJ: IEEE, May 1963).

E. Robertson, ed. *The Solarex Guide to Solar Electricity* (Washington, DC: Solarex Corporation, 1979), pp. 50–52.

A. I. Rosenblatt, *Electronics* (4 April 1974): 99.

P. J. Ruecroft, K. Takahashi, and H. Ullal, *Appl. Phys. Lett.* 25 (1974): 664.

P. J. Ruecroft, K. Takahashi, and H. Ullal, *J. Appl. Phys.* 46 (1975): 5218.

R. P. Ruth, H. M. Manasevit, J. L. Kenty, L. A. Moudy, W. I. Simpson, and J. J. Yang. *Chemical Vapor Deposition Growth,*

ERDA/JPL 954372-76/1, *Q. Report No. 1* (1976).

Ryco Laboratories, Final Report No. AFCRL-66-134, 1965.

P. Sageev, D. Comini, and J. Mortland, *Solar Energy: An Assessment for Business* (Battelle Technical Inputs to Planning Review No. 2, 1979).

G. Sassi, *Solar Energy* 24 (1980): 451.

B. L. Sater and C. Goradia, *Proc. 11th IEEE Photo. Spec. Conf.* (Scottsdale, AZ: IEEE, 1975), p. 356.

D. E. Scaife, *Solar Energy* 25 (1980): 41.

D. G. Schaeler and B. W. Marshall, *Proc. 12th IEEE Photo. Spec. Conf.* (Baton Rouge, LA: IEEE, 1976), p. 661.

R. J. Schwartz and M. D. Lammert, *Proc. IEEE Int. Electron Dev. Mtg.* (Washington, DC: IEEE, 1975), pp. 350, 353.

J. C. Schwartz, T. Surek, and B. Chalmers, *J. Electronic Mat.* 4 (1975): 225.

Science Applications, Inc., *Characterization and Assessment of Potential European and Japanese Competition in Photovoltaics* (Golden, CO: Solar Energy Research Institute, DOE Subcontract No. XP-9-3251-1, 1979), p. 3–15.

R. Scott, *The Solar Market, Proc. Symp. on Competition in the Solar Energy Industry* (Washington, DC: Federal Trade Commission, June 1978) p. 295.

Sensor Technology, Inc., *Data Sheet Number 185* (Chatsworth, CA, 1980).

B. D. Shafer, E. C. Boes, and D. G. Shueler, *Proc. DOE Annual Photovoltaics Program Review for Technology and Market Development* (Hyannis, MA: DOE, 1980), pp. 29–59.

P. Shah, *Solid State Electron.* 18 (1975): 1099.

R. F. Shaw and A. K. Ghosh, *Solar Cells* 1 (1980): 431.

J. L. Shay, S. Wagner, K. J. Bachmann, E. Buehler, and H. M.

Kasper, *Proc. 11th IEEE Photo. Spec. Conf.* (Scottsdale, AZ: IEEE, 1975), p. 503.

J. L. Shay, S. Wagner, M. Bettini, K. J. Bachmann, and E. Buehler, *IEEE Trans. Elec. Dev.* ED-24 (1977): 483.

Shell Oil Company, *The National Energy Outlook* (March 1973).

J. Shewchan, *Annual Progress Report* (Solar Energy Research Institute, 1980), Contract XS-9-8233-1.

L. Shiozawa, G. Sullivan, and F. Augustine, *Proc. 7th IEEE Photo. Spec. Conf.* (Pasadena, CA: IEEE, 1968), p. 22.

W. Shockley and H. Wueisser, *J. Appl. Phys.* 32 (1961): 510.

R. Singh and J. D. Leslie, *Solar Energy* 24 (1980): 589.

D. R. Smith, *Proc. DOE Ann. Photo. Program Rev. for Tech. and Market Dev.* (Hyannis, MA: DOE, 1980), p. 271.

Solar Photovoltaic Energy Research, Development, and Demonstration Act of 1978. Public Law 95-590, Section 11a.

R. J. Soukup, *J. Appl. Phys.* 48 (1977): 440.

A. Spakowski and A. Forestieri, *Proc. 7th IEEE Photo. Spec. Conf.* (Pasadena, CA: IEEE, 1968), 155.

W. E. Spear and P. G. LeComber, *Solid State Commun.* 17 (1975): 1193.

W. E. Spear, P. G. LeComber, S. Kalbitzer, and G. Muller, *Philos. Mag. B* 39 (1979): 159.

Spectrolab, Inc., *Photovoltaic Systems Concept Study* (Washington, DC: ERDA, 1977).

V. Srinivasan and E. Rabinowitch, *J. Chem. Phys.* 52 (1970): 1165.

A. G. Stanley, "Degradation of CdS Thin Film Solar Cells in Different Environments," *Technical Note 1970–33.* (Lexington, MA: Lincoln Laboratory, MIT, 1970).

A. V. Stepanov, *Bull. Acad. Sci. USSR, Phys. Series* 33 (1969): 1826.

R. M. Swanson and R. N. Bracewell, *Electric Power Research Institute* Report No. ER-478 (February 1977).

J. Tauc, *Photo and Thermoelectric Effects in Semiconductors* (New York: Pergamon Press, 1972), p. 18.

T. S. teVelde and J. Dieleman, *Philips Res. Rep.* 28 (1973): 573.

Texas Instruments Corp., *First Quarter Report*, 1980, p. 14.

P. Theobald, S. Schweinfurth, and D. Duncan, *Energy Resources of the United States* (Washington, DC: Geological Survey Circular 650, 1972).

A. P. Thomas and M. P. Thekaekara, *Proc. Joint Conf. Am. Sect. ISES and SES Can. Inc. Winnipeg* 1 (August 1976): 338.

W. G. Thompson and R. L. Anderson, *Solid State Electron.* (1978): 603.

E. H. Thorndike, *Energy and Environment: A Primer for Scientists and Engineers* (Reading, MA: Addison-Wesley, 1976), p. 273.

US Department of the Interior, *United States Energy through the Year 2000* (December 1972).

M. P. Vecchi, *Solar Energy* 22 (1979): 383.

P. Viktorovitch, G. Kamarinos, and P. Even, *Prog. 12th IEEE Photo. Spec. Conf.* (Baton Rouge, LA: IEEE, 1976), p. 870.

L. Waller, *Electronics* (31 January 1980): 40.

L. Waller, *Electronics* (28 August 1980): 41.

W. J. Walsh, *Physics Today* (June 1980); 34.

J. C. Wang, R. F. Wood, and P. P. Pronko, *Appl. Phys. Lett.* 33 (1978): 455.

T. A. Weiss and G.O.G. Löf, *Solar Energy* 24 (1980): 287.

Westinghouse Electric Corp., *Conceptual Design and Analysis of Photovoltaic Systems* (Washington, DC: ERDA, 1977).

R. Whitaker and J. Birk, *EPRI Journal* 8 (October 1976): 6.

C. W. White, J. Narayan, and R. T. Young, *Science* 204 (1979): 4l61.

D. C. White, Energy Laboratory, *Final Report Submitted to the National Science Foundation—Dynamics of Energy Systems* (Cambridge, MA: MIT, 1973).

E. W. Williams, K. Jones, A. J. Griffiths, D. J. Roughley, J. M. Bell, J. H. Steven, M. J. Huson, M. Rhodes, and T. Costich, *Solar Cells* 1 (1980):357.

M. Wolf, "Limitations and Possibilities for Improvements of Photovoltaic Solar Energy Converters," *Proc. IRE* 48 (1960): 1246.

M. Wolf, *Energy Conv.* 11 (1971): 63.

R. F. Wood, R. T. Young, R. D. Westbrook, J. Narayan, J. W. Cleland, and W. H. Christie, *Solar Cells* 1 (1980): 381.

J. M. Woodall and H. J. Hovel, *Appl. Phys. Lett.* 30 (1977): 492.

M. S. Wrighton, *Technology Review* (May 1977): 30.

M. S. Wrighton, A. B. Ellis, P. T. Wolczanski, D. L. Morse, H. B. Abrahamson, and D. S. Ginley, *J. Am. Chem. Soc.* 98 (1976): 2774.

J. Wysocki, *Solar Energy* 6 (1962): 104.

J. Wysocki and P. Rappaport, *J. Appl. Phys.* 31 (1960): 571.

K. Yamaguchi, H. Matsumoto, N. Nakayama, and S. Ikegami, *Jap. J. Appl. Phys.* 15 (1976): 1575.

P. J. Zanzucchi, C. R. Wronski, and D. E. Carlson, *J. Appl. Phys.* 48 (1977): 5227.

J. A. Zoutendyk, *Solar Energy* 30 (1980): 249.

Index

796942